The Making of Federal Coal Policy

***Related* Duke Press Policy Studies**

The World Tin Market

Political Pricing and Economic Competition

William L. Baldwin

The Politics of Hazardous Waste Management

Edited by James P. Lester *and* Ann O'M. Bowman

Global Deforestation and the Nineteenth-Century World Economy

Edited by Richard P. Tucker *and* J. F. Richards

What Role for Government?

Lessons from Policy Research

Edited by Richard J. Zeckhauser *and* Derek Leebaert

The Politics of Canadian Airport Development

Lessons for Federalism

Elliot J. Feldman *and* Jerome Milch

The Making of Federal Coal Policy

Robert H. Nelson

Duke Press Policy Studies
Duke University Press Durham, N. C. 1983

Printed in the United States of America
on acid-free paper

Library of Congress Cataloging in Publication Data

Nelson, Robert H. (Robert Henry), 1944–
The making of federal coal policy.

(Duke Press policy studies)
Includes bibliographical references and index.
1. Coal trade—Government policy—United States.
2. Coal trade—Government policy—United States—History. I. Title. II. Series.
HD9546.N44 1983 333.8′22′0973 83-7155
ISBN 0-8223-0497-X

For my parents, Robert A. and Irene R. Nelson

Contents

Preface

This book is a study of how the policies for managing coal owned by the federal government have been formed. The book focuses on the period from February 1973 to January 1981. During this period new federal coal leasing was suspended, and the federal government was engaged in the development of a new program for managing federal coal. Federal coal is one of the most important publicly owned resources. Production of federal coal in 1981 was 14 percent of total U.S. coal production; federally owned reserves—mostly in the West—constitute around a third of the total coal reserves of the United States.

Aside from the importance to the nation of federal coal, I had a special reason for choosing to write about federal coal policy. In 1975 I joined the Office of Policy Analysis in the Department of the Interior, the agency responsible for managing federal coal.[1] As a member of the office, a considerable part of my time for the next six years was spent studying federal coal policies. During much of 1978 and 1979 I was effectively detailed to the Office of Coal Leasing Planning and Coordination, a small Interior group with the chief responsibility for putting together and implementing a new federal coal-leasing program.[2] In short, I already knew a great deal about federal coal management and policy before beginning the book.

My role as a direct participant has obvious advantages and disadvantages. On the plus side, I can describe the development of federal coal policy from close, firsthand knowledge. I also have had wide access to internal memoranda, studies, papers, and other background materials that an outside researcher would in many cases have difficulty in locating. All but a few official records of government decisions often disappear with surprising speed. Finally, the experience of studying and analyzing federal coal issues for a number of years has exposed me to every point of view. In addition to outside views, I began work in the Ford administration, continued through the Carter administration, and now am again involved with federal coal policy in the Reagan administration. (As of this writing, I am still a member of the Interior Office of Policy Analysis.)

On the other hand, my close personal involvement with the federal coal program has the obvious potential liability that I may have been too close to the subject to see its broader significance. I may have been biased by past positions taken in internal debates or by commitments to existing policies. Recognizing these possibilities, I have conciously tried to put myself in the frame of mind and to approach the subject as if I were an outsider coming in to examine a historical record. I find that I am now critical of some actions and policies I previously defended.

My task has been easier because the study of federal coal leasing is not my first experience in taking a broad look at an important area of public policy. This book, in fact, contains several key themes also found in a previous book on land-use zoning.[3] In hindsight, I have been struck by the number of features common to zoning and federal coal management. Both of these government activities are products of the progressive era. Zoning was first adopted in the United States in 1916, while the Mineral Leasing Act setting the foundations for federal coal management was enacted in 1920. In both cases, however, progressive ideals bore little relation to the actual practice. Yet, because even today progressive principles still constitute the formal foundations, these principles have lived on as myths and fictions often repeated but seldom actually followed.

The central problem in both zoning and federal coal management was the progressive conviction that scientific methods could be applied to social and administrative problems with sufficient authority that they would generate a social consensus—thereby eliminating the necessity for a political resolution. The progressive movement assumed that it would be possible to divorce much of government from politics and political "interference"—leaving these responsibilities instead to "experts." Formal planning plays a critical role in progressive thinking as the vehicle by which the conclusions of the experts are developed, officially communicated to the public, and then used to control government actions. However, as it has become evident that technical expertise cannot achieve the results expected, the basic structure of progressive thought and institutions has been undercut. We are still struggling to find a replacement, as the histories of zoning and federal coal management both illustrate. One popular option—unsatisfactory in my view—is to see government as the achievement of a happy balance among competing interest groups—so-called "interest-group liberalism."[4]

As these remarks suggest, I attempt in this book to assess the lessons offered by the history of federal coal policy for some broader political and economic issues. The book focuses in particular on the surprisingly large role played by ideology in government policy making. The development of the federal coal program is shown to reflect closely a competition among three ideologies—labeled as "conservationism," "planned-market liberalism," and "interest-group liberalism." My analysis lends support to a recent observation of Harvard professor James Q. Wilson concerning "the growing importance for policy making of the ideas of political elites. A political scientist such as myself, trained in the 1950s when politics was seen almost entirely in terms of competing interests, was slow to recognize the change."[5]

Problems of federal coal management described in the book reflect and illustrate a broader crisis in all of public land and resource management. The recent "Sagebrush Rebellion" and other challenges to existing public land institutions are symptoms of this crisis.[6] The more fundamental cause is an erosion of the progressive intellectual foundations on which public land management has rested and the failure to develop a satisfactory replacement. My own suspicion is that these factors will cause major changes in public land institutions sooner than most people now believe likely.

My service at the Interior Department has been partly a personal adventure to examine the workings of government firsthand. I long ago recognized and have become even more convinced of the importance for most policy issues of studying institutional details. Indeed, I have become an advocate of reviving the institutional concerns and attention to practical results that characterized an older tradition of political economy. I am an economist, but it also now seems obvious to me that in their broader aspects politics and economics cannot be separated. I have also come to believe in the critical importance of studying closely the history of policy issues in order to understand current policy debates.

The Interior Department's Office of Policy Analysis partly supported the writing of this book, which began as a review of the development of the federal coal program in the 1970s, written primarily for the purpose of assisting in federal coal-policy formulation. One of the objectives of the many policy-analysis offices created in federal agencies in the 1970s is to provide a longer-run view of policy issues. On completion of this review, it was well received and I was encouraged to make it available for a wider audience. I had full freedom to say what I wished, and the resulting book has not been either approved or reviewed by the Office of Policy Analysis or the Interior Department. Indeed, the examination of wider political-economic subjects probably was not exactly what the Office of Policy Analysis expected. I have written the book with the conviction that the public gains most from interchange among strong viewpoints forthrightly stated.

There are a number of people to whom I owe a great deal for the opportunity to write this book. The former director of the Office of Policy Analysis, Lester Silverman, gave me the time and encouragement to undertake a long-range study of the future of public land management. The focus in this book on federal coal policy emerged later from this undertaking. The director of the economics staff within the Office of Policy Analysis, Robert K. Davis, has been strongly supportive of my writing efforts throughout the study. Donald Sant, who succeeded Lester Silverman as the director of the Office of Policy Analysis, also encouraged my investigations. Finally, Heather Ross, former Deputy Assistant Secretary for Policy, Budget, and Administration from 1977 to 1981, was willing to spare enough of my time from short-term bureaucratic tasks to allow examination of some broader subjects—where the payoff might be high but the risk was also high.

The book has been read by a number of people who have made helpful comments. I want to thank, in particular, Donald Bieniewicz, not only for the comments and encouragement he offered in writing this book, but also my earlier efforts in the field of public land management. I am also grateful to other readers and commentators including Eugene Bardach, David Gulley, Christopher Leman, Charles Rech, Peter Reuter, Dan Tarlock, Jeffrey Wasserman, and Aaron Wildavsky. Joseph Browder and John Leshy read and commented on an earlier draft, much different in many ways from the final version.

I was greatly assisted by the typing skills of Beverly Bergman, Cherie Franklin, and Ella Johnson.

The Making of Federal Coal Policy

Introduction

The 1980 United States Census confirmed what had already been apparent for some time: a major movement of people and production towards the West and South. Seven of the ten most rapidly growing states in the United States from 1970 to 1980 were located in the Rocky Mountain West. Nevada was the single fastest growing state (growing 64 percent); Arizona ranked second (53 percent); Wyoming fourth (42 percent); Utah fifth (38 percent); and Colorado (31 percent), Idaho (32 percent), and New Mexico (28 percent) were also in the top ten.

The high rates of growth of the states in the Rocky Mountain West were caused in part by the small initial base from which their growth jumped. These states had long been the most sparsely populated in the United States. The land was harsh and mostly unproductive without irrigation; it was, in fact, a true desert in many areas.[1] The economy of the Rocky Mountain West had long consisted of ranching, mining, and patches of irrigated farming, along with the services necessary to support these activities. More recently, the attractive climate and natural setting had made recreation a growth industry. In addition, growing numbers of retirees, especially in the Southwest, and others who were willing to sacrifice some income for an outdoor way of living had arrived.

But the unusually rapid population spurt of the 1970s had one main explanation: the Organization of Petroleum Exporting Countries and rising oil and gas prices. The nation discovered that a major share of its still untapped but nevertheless reasonably accessible energy resources were to be found in the Rocky Mountain West. United States uranium production has always been and would continue to be concentrated in these states. If not as widely known, substantial oil and gas had also been produced for many years, especially in New Mexico and Wyoming. The prospects for further large western oil and gas discoveries rose sharply, however, with new discoveries in the "Overthrust Belt," a geographic formation running from Canada to Mexico along which two old continents had drifted together. There are wide hopes to find here new pools of oil conceivably as large as found in Alaska in the 1960s.

Almost all the nation's prime oil shale is concentrated in a small area near the intersection of Colorado, Utah, and Wyoming. There is more than enough oil shale to supply all U.S. oil requirements for decades if not centuries. If environmental problems can be resolved, and the economics ever become favorable, oil shale could eventually provide a secure domestic source for a significant part of U.S. petroleum production.

Western Coal Production

But for the short run, and maybe the long run as well, the most important new source of energy from the Rocky Mountain West will be coal.[2] Earlier in the century small amounts of western coal were used to power steam locomotives. With the introduction of diesels after World War II, however, western coal production dropped off to very minimal levels. Then, in the early 1970s transportation of low-sulfur western coal to new coal-burning power plants in parts of the Midwest became economic. Because of environmental regulations, it had become cheaper to import western coal than to install the expensive scrubbers needed to clean high-sulfur eastern and midwestern coal.

The use of coal for electric power generation did not dramatically increase until after World War II; it was still only 88 million tons—18 percent of U.S. coal use—as recently as 1950. As coal was phased out in its more traditional transportation, industrial, and home heating uses after the war, only the rapid rise in power-plant burning of coal prevented a complete collapse of national coal production. But by 1975 the nation's utilities burned 403 million tons of coal—72 percent of the U.S. coal use. Still, coal had supplied 50 percent of the nation's energy in 1910; in 1975, it was providing only 18 percent.[3] Coal was the one energy source which was available right away and in huge quantities—enough supply in the United States potentially for hundreds of years—and that was much cheaper than foreign oil imports.

Although not widely known until recently, more than 50 percent of the U.S. coal reserves are located west of the Mississippi River. Traditionally, western coal was too costly to transport to attract eastern buyers. There was also little demand for western coal in the West itself. The West obtained ample amounts of much cheaper and cleaner power from oil- and gas-burning power plants and hydroelectric facilities. The expectation had long been that nuclear power would then meet additional long run needs. But environmental opposition to new dams and to nuclear power, combined with the OPEC-spurred escalation in oil and gas prices, suddenly and unexpectedly left coal as the main power plant fuel of the future in the West.

Western coal production, as recently as 1967, was only 29 million tons, just 5 percent of total national production. But by 1972 it had climbed to 60 million tons, and continued to rise rapidly in the 1970s, reaching 251 million tons and 30 percent of national production in 1980. Most forecasts indicate more of the same rapid growth in the 1980s; the Department of Energy has developed projections showing that western coal production will rise to 755 million tons in 1990, supplying 47 percent of the nation's coal production.[4] Most of western coal will be burned in the West, although a significant amount—18 percent in one estimate—will be shipped to midwestern and eastern markets.[5] Besides its low sulfur content, western coal is attractive to these markets because of its very low mining costs, as low as $5 and $10 per ton compared with $20 to $35 per ton for typical eastern coal. More than 90 percent of western coal is expected to be surface mined, much of it from very thick seams lying near the surface, and thus very efficiently mined. Surface mining has risen steadily from 22 percent of total U.S. coal production in 1947 to 59 percent in 1980.

The future of U.S. coal production thus lies towards the West. Of a total growth in U.S. coal production of 839 million tons forecast by DOE to take place in the 1980s, traditional Appalachian sources are expected to contribute only 109 million tons, the Midwest 197 million tons, and areas west of the Mississippi the great bulk of the increase, 533 million tons. Even if actual growth turns out to be less than these DOE projections, as the most recent trends suggest, there will still be a strong western orientation.

The production in the West by 1990 of much larger amounts of coal will transform the character of many areas. The state most affected will by Wyoming, now the least populated state except for Alaska. Coal production in Wyoming could rise from 4 million tons in 1967 to more than 200 million tons in 1990, exceeding 10 percent of national production. Parts of Utah, Montana, New Mexico, and Colorado will also be the location of much coal mining and face consequent infusions of population and income. Continued growth of western coal production will be a main factor helping to sustain the rapid growth of the Rocky Mountain West already seen in the 1970s.

Federal Coal Ownership

Except for Texas, all the coal in the West was once owned by the federal government. But large amounts were included in nineteenth-century railroad grants and in other federal land disposals under the Homestead Act and various other public land laws. Starting around 1910, however, the federal government began to retain the mineral rights, even when it disposed of federal surface ownership. As a result, not much additional federal coal was transferred into private hands. Today, the federal government still owns around 60 percent of the U.S. coal reserves located west of the Mississippi River—constituting around 33 percent of all U.S. reserves.[6]

Moreover, much of the state and privately owned coal in the West can only be developed jointly with federal coal because of closely intermingled coal-ownership patterns. The railroad coal in old checkerboard areas, for example, must be developed in most cases jointly with federal coal. It is estimated that at least 50 percent of the nonfederal coal in the West (20 percent of all western coal) is so intermingled that it should be developed in conjunction with federal coal. The federal government thus controls around 80 percent of all western coal development by virtue of its dominant ownership position.

Furthermore, the federal ownership is highest in some of the most desirable areas in the West for coal mining. The Powder River Basin in northeast Wyoming and southeast Montana is sometimes called the "Kuwait" or "Saudi

Arabia" of United States coal. It contains vast reserves—more than 143 billion tons—in coal seams often close to the surface and in some places 100 or more feet thick (a fifteen-foot seam would be exceptional in the East). Eighty percent of the coal in the Powder River Basin is owned by the federal government. The Interior Department estimates that, except for Indian coal, only 7 percent of the coal in this region could be developed independent of any federal coal availability.

In short, there is not much of a distinction between western coal development and federal coal development. As one goes, so will go the other. Policies for managing federal coal thus will have a major impact on both national energy production and the future development of the Rocky Mountain West. The federal agency bearing this heavy responsibility is the Bureau of Land Management in the Department of the Interior.

A historic event for federal coal management occurred on January 13, 1981: the first sale of federal coal leases in ten years on other than an "emergency" basis. With the enactment in 1920 of the Mineral Leasing Act, the federal government stopped selling coal outright and henceforth would only lease it. Federal coal leasing had proceeded sporadically until the 1960s when it picked up sharply. By 1971 the Interior Department had become concerned about the obvious growing national importance of federal coal in the West, the large amounts of coal under lease, and the fact that federal coal management had thus far received very little attention from top Department officials. In May 1971 the Department informally ceased issuing new coal leases, never imagining that this moratorium would last for ten years. During just the period when western coal was emerging as a key national energy source, and the federal government controlled the production of most western coal, almost no new federal coal was leased. As will be examined in the chapters to follow, once leasing was stopped, it proved impossible for many years to achieve sufficient acceptance of a new federal coal program to be able to resume leasing again.

Two programs for leasing federal coal were stillborn, one preliminarily announced in mid-1974 and a second in early 1976. A third program was installed in 1979, still surrounded by much controversy, although in this case accepted widely enough that leasing could finally start again in 1981. The fierce debates over federal coal leasing concerned, first, the actual amount of coal that should be leased by the federal government and, second, the exact nature of the program by which leasing should be accomplished. The design of the federal coal program raised basic questions about the appropriate relationship of government and the private sector. As will be seen, the answers given tended to reflect the basic ideological outlooks of the respondents.

The term "ideology" has a negative connotation for many Americans who like to think of themselves as "pragmatic." However, an ideology is really just a way of organizing widely disparate information into a coherent framework. Without some such broader understanding and sense of direction, purposeful behavior becomes difficult if not impossible. Even a pragmatic approach is itself an ideology of sorts, although one that denies the existence of any single organizing principle.

An ideology by its nature must involve some simplification of the world. Good ideologies capture the essence of the matter; poor ideologies lead people in the wrong directions. The scholarly proponents of an ideology are more likely to note the inevitable qualifications, omissions, and other limitations. However, in the public policy arena these complications are likely to be neglected or lost altogether. The ideology has its greatest impact on government decisions in its simpler and cleaner versions. In this book one main aim is to analyze the several key ideologies, not as they were refined by leading—often academic—proponents, but as perceived by the direct participants and as they had an actual impact in the public policy-making process.

I. The Conservationist Foundations for Federal Coal Management

Introduction

The size of federal coal holdings in the West creates an unusual situation. The government, as noted, has effective control over most of a major industry, western coal production. It thus has had to face a rare question in the United States: given the authority to plan the bulk of a whole new industry, how should this authority be exercised?

The federal role has perhaps been as big in the design of a few industries, mainly those involving basic transportation and communications infrastructure such as are (or have been) regulated by the Interstate Commerce Commission, Civil Aeronautics Board, and Federal Communications Commission. In the case of the Post Office, the federal government nationalized a whole industry. It has also built and operated regional power systems such as the TVA and western hydroelectric and irrigation projects. But the government has never directly undertaken to control the development of basic industries such as steel, automobiles, electronics, oil and gas, or, until recently, coal. Even in those cases where fairly tight federal control has been found, it has typically been a process of incremental imposition of controls on a previously existing industry. Taking into account the lack of previous close federal supervision of basic industries, and the opportunity to start virtually from scratch, there have been few, if any, U.S. precedents for the circumstances of federal coal management.

One option would have been to seek to renounce those responsibilities and turn the whole matter over to the private sector. The federal government could have endeavored to sell off its coal holdings in order to create an ordinary private market in western coal. Another option would have been for the government to directly produce federal coal itself, in effect nationalizing much of the western coal industry. There were also many intermediate possibilities.

Although the federal government has been committed historically to a leasing system, this commitment affected these issues less than might be thought. The government could give the holder of a federal lease a free hand in making all the decisions concerning future production of leased coal—for practical purposes just as if he owned the coal. Competitive bidding could determine the price of coal under either outright sale or leasing. Provision could also be made under either system for future royalty payments to the federal government. Finally, if the length of the lease were indefinite as well, the practical distinction between selling off federal coal and leasing this coal would virtually disappear.

On the other hand, a leasing system can also approximate closely the results of direct federal production of the coal. Under leasing the federal government, if it wishes, can still retain almost complete control over the rate and amount of production from federal leases. The lessee might be controlled to such an extent that he becomes little more than the producing agent of the federal government. Once the federal government took its royalties and rentals, competitive bidding among coal companies for leases would further reduce net revenues accruing to winning companies to a minimal payment for production services. In short, leasing versus the outright sale of federal coal has been important for its symbolism, evoking images of "public" versus "private" resource ownership, but in itself it does not resolve the basic management questions.

About the only unavoidable consequence of a leasing policy is to commit the government to hold at least some of its coal resources (otherwise, there is nothing to lease). However, the Mineral Leasing Act of 1920 went far beyond this minimal obligation. It not only formally ended the era of federal coal disposal, but generally sought to implement an ideology requiring close government planning and supervision of federal coal production. This ideology was conservationism, the application in the special field of natural resources of the broader ideology of the progressive movement.

1. Conservationism and the Mineral Leasing Act of 1920

The management of federal coal is part of the history of the public lands—indeed, a generally obscure part until the 1970s. During the nineteenth century the policies to dispose of the public lands were among the central public issues of the day. One recent study commented that "the social, political and economic life of the United States was surrounded by questions of the public domain" during the nineteenth century.[1] The Homestead Act, the railroad land grants, the range wars, and other episodes involving the public lands are now part of American folklore. The early laws for federal coal were mainly an adaptation of the general land laws to the special circumstances of federal coal.

The Coal Land Laws

Coal received no special treatment in the public land laws until the 1860s; until then, if someone needed coal from public lands, he simply acquired the land containing it and received the coal rights as well. But in 1864 Congress enacted the first legislation dealing specifically with federal coal resources. Reflecting a concern that federal lands already known to possess coal were worth much more than ordinary land, the legislation provided for sale of coal lands at public auction for a minimum price of $20 per acre (at that time the government sold ordinary land for a minimum of $1.25 per acre). In 1865 Congress modified the law to require that purchasers be miners and limited the acreage acquired to 160 acres. The Coal Lands Act of 1873 established a minimum price related to the distance from a railroad—$20 per acre for coal lands within fifteen miles and $10 beyond; it also provided for purchase of up to 640 acres by qualified associations. The 1873 Act subsequently controlled the sale of federal coal for almost fifty years until the Mineral Leasing Act of 1920.[2]

Unlike coal, most minerals on the federal lands were disposed of under the location and patent system established by the Mining Law of 1872, which today is still the basic legislation for gold, silver, copper, and other "hardrock" minerals. Unless the land is withdrawn, prospectors are free to enter federal lands to explore for these minerals, file a claim if anything appears promising, and obtain a patent (ownership) if they can demonstrate the presence of a sufficiently valuable mineral deposit. The mining laws were designed to create the incentive to explore for minerals that were usually very hard to find. However, huge amounts of coal were already known to lie in continuous beds that sometimes could even be identified from visible surface outcroppings. Where coal was already known to be present, even if in uncertain amounts, it made more sense for the government to sell it, rather than give it away without charge. As a result, there was much less need to create an exploration incentive, and federal coal was sold directly, instead of being made available under the usual mining laws.

The coal laws proved to be a limited example of why almost all the public land laws of the time encountered great difficulties and often failed in their stated purposes. The attempts to limit unrealistically the size of coal lands that could be acquired were one main source of the problem. The mine owner could not be confident he would be able to acquire enough coal to meet his mining requirements. As one observer put it, "The coal lands . . . provide for maximum entries hardly sufficient to establish a profitable mine."[3] Equally troublesome, it was much more expensive to acquire lands under the coal laws than under the general land laws. The lines between coal lands and other lands were not always well defined. Hence, the familiar practice of the time, wide evasion of the public land laws, resulted in this case as well. Multiple entries were often filed to obtain larger coal acreages than legally permitted, and lands with valuable coal were improperly acquired under the Homestead and other laws intended for grazing and agricultural lands. Even on lands sold under the coal laws, the government failed to receive full value; the price was much less than going prices for comparable private coal properties.

The railroads were the main consumers of western coal and not that much was needed for their use. Only 400,000 acres were sold under the coal laws in the first thirty-four years. Hence, the abuses of the coal laws were less conspicuous than other public land laws and for a long time received little attention. However, the conservation movement during the Theodore Roosevelt administration brought to light abuses that led Roosevelt in 1906 to withdraw a huge area of 66 million acres from coal entry (although not as large as his withdrawals of 148 million acres for the national forest system). The massive removal of federal coal from availability for use—along with similar Roosevelt actions with respect to other public land resources—stirred bitter resentment in the West, which contested their legality. By 1908, however, more than half the coal lands, 38 million acres, had been restored to the public domain; the hasty withdrawals in 1906 had in fact included many lands actually having little if any coal.

But the large areas remaining withdrawn, in addition to all the other lands that might be identified to have coal reserves and thus required to be sold at a much higher price, threatened to put a crimp in the Homestead and other general land disposal laws. To resolve this problem, Congress adopted a new policy of separating surface and mineral ownership. A 1909 act allowed those who had previously entered coal lands to acquire surface ownership with the coal rights reserved to the federal government. Legislation in 1910 permitted ordinary entry, under the Homestead and other land laws, of areas previously withdrawn or classified as coal lands, again with the rights to federal coal retained. In 1912 Congress provided for state selections and certain sales limited to surface rights. Finally, the Stock-Raising Homestead Act of 1916 had the widest impact; henceforth, it reserved all mineral rights to the federal government on land acquired under the act.

As a result of these laws a widespread separation of surface and coal ownership now exists; according to one study of the prime coal lands in the West, only 47 percent of federally owned coal also has federally owned surface rights.[4] In the coal-rich Powder River region only about 23 percent of the federal coal also has federal surface ownership. The Powder River region had fairly productive lands and much homesteading occurred there in the early part of the century. Now that large-scale federal coal development is finally occurring, the Interior Department today must contend with the difficult problems caused by the separation of surface and mineral ownership.

Progressivism and the Conservation Movement

After 1906 the withdrawal of federal coal lands raised a number of basic questions. Should the federal government simply clean up the abuses and then go back to selling coal resources whenever and wherever anyone asked to buy them? Should the government instead seek to limit its sales of coal to the amounts needed right away for actual coal production? What should be the role for speculative purchasers of coal? Should a leasing system instead be adopted? What federal controls should be maintained over production of federal coal? At the turn of the century the federal government was sharply revising the old disposal policies for other public lands; for example, by 1908 it had created most of the present system of national forests under Forest Service management. Should a similar revision of disposal policies be undertaken with respect to federal coal as well?

The answers to such questions were in large part provided by the conservation movement. Indeed, the ideology of conservationism underlay the establishment of the Bureau of Reclamation in 1902, the creation of the Forest Service in 1905, and other major changes in public land policy that occurred from 1890 to 1920. Conservationism was one of the principal themes of the administration of President Theodore Roosevelt. As a result of conservationist influence, the basic philosophy of public land management shifted from disposal to retention during this period. In form at least, the public lands today are still being managed under the institutional framework and guiding principles of the conservationist ideology. To be sure, the actual nature of public land management has changed significantly since the turn of the century, even while the formal appearance has been much less altered.

As noted above, the conservation movement was itself a central part of the broader progressive movement. Progressivism arose from a wide discontent with many results of an unfettered individualism in nineteenth-century America, sanctioned under the free-market principles of classical liberalism. The classical liberal ideology offered a vision of rapid social and economic progress, wide individual opportunity for all, and wide personal liberty—achieved in economic affairs by means of the free market and in political affairs by democratic government. Nevertheless, as the historian Richard Hofstadter observes, a crisis of confidence in the existing system had developed by the turn of the century:

> The world with which the majority even of the reformers was most affectionately familiar was the passing world of individual enterprise, predominately small or modest-sized business, and a decentralized, not too highly organized life. In the Progressive era the life of business, and to some degree even of government, was beginning to pass from an individualistic form toward one demanding industrial discipline and engendering a managerial and bureaucratic outlook. . . . Most Americans who came from the Yankee-Protestant environment, whether they were reformers or conservatives, wanted economic success to continue to be related to personal character, wanted the economic system not merely to be a system for the production of sufficient goods and services but to be an effectual system of incentives and rewards. The great corporation, the crass plutocrat, the calculating political boss, all seemed to defy these desires. Success in the great corporation seemed to have a very dubious relation to character and enterprise; and when one observed the behavior of the plutocracy, it seemed to be inversely related to civic responsibility and personal restraint. The competitive process seemed to be drying up. All of society was felt to be threatened—not by economic breakdown but by moral and social degeneration and the eclipse of democratic institutions.[5]

In these circumstances Americans saw a necessity to turn to government to bring greater control over the new order. New social principles, however, were necessary; the new government role could not be reconciled with traditional nineteenth-century ideology. In a famous 1887 article Woodrow Wilson, then a rising young professor, set out several key themes of a newly emerging ideological foundation, progressivism. He explained the critical distinction that was to underlie progressive thinking—a division of the governing process into two separate spheres, ordinary politics and scientific administration. Wilson considered that "the field of administration . . . is removed from the hurry and strife of politics."[6] The political realm would maintain the traditions of American democratic government; the administrative realm, which would be large, could be entrusted to technical experts in management removed from politics. These experts would apply modern scientific methods to introduce an efficiency in the administration of government that would match the efficiency that new technologies and skillful planning were demonstrating throughout the world of business. Democratic politics would set the broad goals, and the new corps of professional administrators would achieve them with maximum efficiency. As a leading student of progressive ideology commented in 1948:

> Every era . . . has a few words that epitomize its world-view and that are fixed points by which all else can be measured. In the Middle Ages they were such words as faith, grace and God; in the eighteenth century they were such words as reason, nature, and rights; during the past fifty years in America they have been such words as cause, reaction, scientific, expert, progress—and efficient. . . .
>
> However natural, it is yet amazing what a position of dominance "efficiency" assumed, how it waxed until it had assimilated or overshadowed other values, how men and events came to be degraded or exalted according to what was assumed to be its dictate. It became a movement, a motif of Progressivism, a "Gospel."[7]

Progressives had a great faith in the benefits of public planning; it was to be the scientific instrument for the achievement of efficiency and other progressive objectives: "Planning is the means by which the discipline of Science applied to human affairs will enable man to incarnate his purposes. It is the inevitable link between means and ends. Moreover, it is in itself an inspiring ideal." To provide for such planning, considerable expansion of the government role would be necessary. But progressives were not concerned about this prospect: "The Good Life will be planned—but to what extent? . . . Almost without exception they look favorably upon government, regard it as a desirable instrument for the accomplishment of individual and community purposes, profess indifference or express favor at proposals to extend the range of its operation or control." How are we to be assured that the power of an expanding government will actually serve the general interest? Here the progressive belief in science becomes most crucial. The answer "to the problem of the basis of decision is that 'science,' 'facts,' 'measurement' answer questions of 'what to do?' It asserts that what is objective can and should 'determine,' that the imperative of 'the facts' should be substituted for chance and will." To discover the scientific truth in the facts of the situation, progressives considered that a new set of scientifically trained public administrators would be required: "Prominent in the early writings is insistence upon the necessity for 'experts' in government service."[8]

The Federal Reserve System—established in 1913—is the quintessential existing embodiment of progressive principles such as administrative independence from democratic politics. Although it seems a curious concept today, perhaps the best way to grasp the true progressive vision is to imagine a large number of organizations like the Federal Reserve System administering almost all matters for which government is currently responsible. The creation of the Interstate Commerce Commission, Federal Trade Commission, and other independent regulatory agencies was also in part a reflection of progressive ideology.

The conservation movement took the basic tenets of progressivism and applied them to the special area of natural resources. The leading conservationist of his day, Gifford Pinchot, advocated government management of natural resources because it would be the quickest way to obtain the efficiencies of true scientific management. Atomistic producers might lay waste to forests from lack of knowledge or incentive to apply expert forestry methods; federal ownership would instead bring modern science and the latest planning techniques. The historian Samuel Hays writes that the conservation movement preached a "gospel of efficiency" based on applied science: "Conservation, above all, was a scientific movement, and its role in history arises from the implications of science and technology in modern society." Moreover, following progressive tenets, politics clearly had little place: "Since resource matters were basically technical in nature, conservationists argued, technicians, rather than legislators, should deal with them." Instead of the traditional patronage and generally lax management of the political appointees of the time, the new expert administrators would employ disinterested, "rational planning to promote efficient development and use of all natural resources." As Hays sums up conservationism:

> The broader significance of the conservation movement stemmed from the role it played in the transformation of a decentralized, nontechnical, loosely organized society, where waste and inefficiency ran rampant, into a highly organized, technical, and centrally planned and directed social organization which could meet a complex world with efficiency and purpose. This spirit of efficiency appeared in many realms of American life, in the professional engineering societies, among forward-looking industrial management leaders, and in municipal government reform, as well as in the resource management concepts of Theodore Roosevelt. The possibilities of applying scientific and technical principles to resource development fired federal officials with enthusiasm for the future and imbued all in the conservation movement with a kindred spirit. These goals required public management, of the nation's streams because private enterprise could not afford to undertake it, of the Western lands to adjust one resource use to another. They also required new administrative methods, utilizing to the fullest extent the latest scientific knowledge and expert, disinterested personnel. This was the gospel of efficiency—efficiency which could be realized only through planning, foresight, and conscious purpose.[9]

Dry scientific efficiency, however, could never have aroused popular passions as the conservation movement did. Hays himself saw the importance of other elements: "Roosevelt's emphasis on applied science and his conception of the good society as the classless agrarian society were contradictory trends of thought."[10] Indeed, Hofstadter shifts the emphasis in his interpretation of progressivism. As indicated above, he regards it as essentially an attempt to revive the nineteenth-century individualistic values of American society. The small entrepreneur and the traditional democratic process were threatened by the rise of the giant corporation, the spread of great congested cities, and the general rapid pace of change in modern life. Paradoxically, these developments were driven by the very science that progressivism exalted; the ability of the large corporation to apply planning and scientific management was in fact a main advantage over its atomistic competitors. In his trust-busting and certain other activities, Roosevelt seemed to be trying to restore a market form of economic organization that was losing out to the scientific management that he prized so highly.

The actual behavior of Pinchot's creation, the U.S. Forest Service, supports the Hofstadter at least as much as the Hays interpretation. The Forest Service exhibited a very high morale based on a moralistic fervor for doing good in the world—a revivalist appeal to the good old values. As a former Forest Service chief put it, the agency was "born in controversy and baptized with the holy water of reform."[11] Where science and orthodoxy conflicted, preserving the faith often won out: "While the conservation movement invoked the authority of science, it also resorted to highly emotional appeals in an attempt to enlist support for its policies. Unfortunately, the evangelistic approach introduced elements of parochialism and rigidity into analyses of problems, thereby contributing to erosion of the movement's scientific base."[12] Pinchot himself constantly preached that the Forest Service was doing "good work in a good cause." Its opposition was the "vast power, pecuniary and political" of the "railroads, the stock interests, mining interests, water-power interests, and most of the big timber interests." But Pinchot was sure that virtue and justice would win out: "In the long run our purpose was too obviously right to be defeated."[13]

Full application of conservationist precepts to the federal coal holdings would leave the private sector with little to do. First of all, the government would retain the coal, assuming the basic management responsibility. The government would plan scientifically the location, timing, rate, and type of federal coal development. As a concession to American traditions of private enterprise, the coal might be leased rather than directly produced by the government. However, coal companies would mainly carry out details specified in lease terms and subsequent government instructions. The government would make its own estimates of the amount of federal coal needed for near-term production and would then lease only this amount. Coal industry or other special interest pressures would be considered improper influences on leasing decisions. As far as possible, politics should be excluded; decisions would be based on the studies of economists and other technical experts especially knowledgeable about national energy requirements, the demands for federal coal, and the necessary protection of the environment from coal-mining impacts. The overall theme would be comprehensive federal planning of federal coal development for the public good. As progressives put it, "The natural resources of the nation must be promptly developed and generously used to supply the people's needs, but we cannot safely allow them to be wasted, exploited, monopolized or controlled against the public good."[14]

Such a policy would, of course, be a very radical change from the nineteenth-century policy of simply selling federal coal to anyone who came along asking to purchase it. But progressive and conservationist ideology had captured the mood of the American public in the early part of this century. There was widespread public optimism in all fields that an era of unprecedented human progress and well-being lay ahead, based largely on the powers of science and technical expertise. This mood had real consequences; indeed, its manifestation in conservationism shaped the Mineral Leasing Act of 1920. To be sure, as will be examined in chapter 2, it was much easier to pass legislation symbolic of conservationist aspirations than to radically shift federal coal policy as the new statute directed. Shifts in public attitudes, reflected in rapid rises or declines of ideologies, are often much wider than the changes in actual practice on the ground.

Passage of the Mineral Leasing Act of 1920

The initial efforts to pass a federal leasing law for oil and gas, coal, and several other "leasable" minerals began in the Theodore Roosevelt administration. Following his withdrawals to prevent further coal sales to the private sector, Roosevelt then proposed a leasing system in 1906, followed by a message to the Congress in 1907 saying that "the Nation should retain its title to its fuel resources, and its right to supervise their development in the interest of the public as a whole."[15] However, vigorous opposition to leasing from westerners, combined with lackluster support from eastern progressives who sought even stronger reform measures, combined to defeat his efforts. Westerners feared that federal leasing would exact revenues from the West to benefit national taxpayers, and might limit the availability of federal coal to meet western needs.

The stalemate over federal coal policy persisted for many years, leaving 44 million acres of federal coal lands still unavailable in 1919 for coal production in any fashion—a forerunner to a similar stalemate in the 1970s. But over the years westerners were gradually coming to believe that they might have to accept a leasing law, although only "because we can not get a better policy."[16] One westerner stated in 1919 that, despite his dislike of the proposed leasing legislation, "I may hold my nose and vote for it."[17] Western congressmen in some cases resented the fact that the federal mineral estate could "never pass into private ownership" where it could be taxed; by contrast, easterners were "living in states unburdened by such conditions, states which were once public-land states also, but whose lands have passed entirely into private ownership."[18] A number of western congressmen preferred management of mineral resources by the western states themselves—also a theme of the recent "Sagebrush Rebellion." Nevertheless, by 1920 enough westerners were ready to give grudging support to a leasing system as the only feasible way to move ahead with the desired development of coal, oil, and other mineral resources of the West.

The Mineral Leasing Act thus finally passed in 1920, the basic legislation for management of oil and gas, coal, oil shale, sodium, phosphates, and certain other minerals that the Congress determined should be leased. The conference report on the bill said that "for many years the West has been practically tied up and no development had [occurred], due to the antiquated land laws that have been on the statute books and the strict interpretation of them, which has rendered them valueless, unworkable, and utterly impracticable." The new law was intended to promote "development of the West and of the natural resources of the West and at the same time reserve to the government the right to supervise, control, and regulate the same, and prevent monopoly and waste and other lax methods that have grown up in the administration of our public-land laws." In addition, "Royalties and rentals are provided, so that the government may not be passing to title the natural resources without receiving something in return therefor."[19]

The Mineral Leasing Act provided for two main types of leasing. Under the system for "preference-right" leasing, an applicant could file for a two-year prospecting permit in areas where the government did not know the extent to which coal was actually to be found. In such areas further exploration was considered desirable and the government was prepared to reward it. Hence, if the prospecting permit were granted, and significant deposits of coal were in fact then found, the holder of the permit could apply for a preference-right lease. To obtain the lease the applicant had to demonstrate that coal had been discovered in "commercial quantities," a term whose definition has since stirred wide debate. On such a demonstration, the lease would be issued at no charge.

Where coal was already known to be present, however, leases were to be awarded under a much different leasing system. The government could directly sell such leases, normally through competitive bidding, but conceivably through other suitable means.

The Mineral Leasing Act also established limits on the total lease acreage that could be held by any one party in the same state (since 1964 the limit has been 46,080 acres). Rentals building up to at least $1 per acre from the sixth year onward and royalties of at least 5 cents per ton were mandated. Leases could be renegotiated by the Secretary of the Interior every twenty years. The Act also required "diligent development and continued operation of the mine or mines," although no specific minimum time to begin production was set. However, "payment of an annual advance royalty" could substitute for the maintenance of continuous operations.

A Mandate for Central Planning

The Mineral Leasing Act of 1920 (later modified by the Federal Coal Leasing Amendments Act of 1976), gives the Department of the Interior the authority to control almost all aspects of the development of federal coal. The Department can determine which particular coal deposits will be developed by its decisions to lease or not to lease. Through "diligent development" and "continous operation" requirements it can decide the time by which coal mining must begin and the subsequent rate at which it takes place. A statutory requirement for "maximum economic recovery" gives the Department control over the specific coal seams and the total amount of coal mined from the lease. Environmental controls and a general requirement for approval of the mining plan give the Department substantial influence over the kind of mining operation undertaken. The Interior Department even has the statutory authority, although it has never exercised it, to set the price of coal mined from federal leases. When these government powers are considered together, over all federal coal leases, the Interior Department has the authority to control precisely the total amount of federal coal produced, the geographic pattern of its production, the rate and manner of production, the price of the coal output, and almost any other matter it may decide.

In one sense there is nothing unusual about these powers. All of them are standard rights of property ownership; if a corporation or other private party instead owned the federal coal resources, it could impose precisely the same requirements on their development. Nevertheless, as noted previously, the huge size of federal coal holdings gives the federal government wide control over a whole industry—western coal production. Under the provisions of the Mineral Leasing Act, the federal government thus has the authority to plan the development of this industry. Moreover, the intent of the Mineral Leasing Act—reflecting its conservationist foundations—is that this authority should be exercised. Having precluded private market determination of federal coal development, there is no other way to make all these decisions for a whole industry than by a major exercise in central planning.

2. Congressional Intent and Opposite Results

The critical role played by conservationist ideology was by no means the only respect in which the history of federal coal management closely followed the broader history of public land management. Throughout the long history of the public lands there often has been a wide gap—in some periods really a chasm—between congressional intent as stated in the law and the actual happenings on the public lands. The same fate was to characterize coal management under the Mineral Leasing Act of 1920, along with most of the other public land products of conservationism.

A Long History of Unworkable Public Land Laws

At the end of the eighteenth century Congress adopted the policy to sell the federal lands to raise revenues for the government. Such sales were to be an important revenue source for a new nation still burdened with a large debt from the Revolutionary War. The only other major source of revenue was tariffs on foreign imports. The original thirteen states considered the public lands in the West to be part of the national government's wealth and did not want to give them away—as they saw it, to forfeit their property rights.[1]

But the attempt to implement this policy led to one failure after another. From 1800 to 1820 the federal government extended advance credit to settlers, but, finding itself often unable to collect, then passed a long series of laws delaying and eventually forgiving settler debts. After 1820 the granting of credit was abolished, but this policy simply led to widespread unauthorized entry. When the federal government sought to sell public land which was already occupied by a squatter, organizations of local settlers would often threaten other prospective bidders—with violence, if necessary. Such intimidation eventually became so widespread that informal property rights of squatters to buy their land for the minimum purchase price were effectively established. Congress was pressured into passing a number of laws retroactively confirming the rights of earlier squatters to acquire title to their land for this price. Finally bowing to the inevitable, Congress enacted the Preemption Act of 1841, granting future settlers, as well, the rights to buy up to 160 acres of public land they might choose to occupy. Various measures in the 1840s and 1850s then had the practical effect of gradually reducing the price the government received in land sales. Eventually, in 1862 the Homestead Act simply made it all official; free land would be given in amounts up to 160 acres to settlers who stayed for five years.

The Homestead Act showed that Congress had finally learned from its past mistakes—however slowly. Unfortunately, it also showed that Congress still did not know how to look toward the future. The Homested Act was based on the experience of public land disposal in the rich farmlands of the Midwest. Here, 160 acres had been ample for a family farm. As settlement moved further west, however, it reached areas where there was little rainfall and the land was much less productive. In the Rocky Mountain West, it was usually impossible to farm 160 acres without irrigation. Indeed, several thousand acres actually were necessary in many places for even a modest cattle ranch. The Homestead Act thus never worked in the West, nor did most of the other public land legislation of the next few decades. The huge areas of public lands still in public ownership in the West today are primarily a result of the failures of laws designed to fit circumstances that never existed where they were actually applied.

The failures of public land policy in the second half of the nineteenth century were due most of all to public unwillingness to abandon illusions about the feasibility of wide settlement in the West by small farmers. Because of this refusal to face reality, the Homestead and other public land laws were beset by a central contradiction. They sought to dispose lands rapidly for agricultural development, yet imposed conditions that made such disposal and development very difficult. Captured by the huge success and resulting mystique of the small midwestern farmer, Congress would not hear that cattle and sheep ranching were the only practical uses for most of the land in the mountain West. As a consequence, the only way to get around unworkble land laws was to engage in fraud and extralegal activities, which of necessity became a way of life for many westerners acquiring land in the late nineteenth century. Reform attempts to curtail such activities only aggravated the situation; as one historian put it, they "eliminated much of the flexibility that had enabled persons in the High Plains to acquire control over if not ownership of 320 or 480 acres." Thus "Congress made it even more necessary for ranchers and others seeking to gain ownershp of economic units to resort to fraud in a more systematic way than they had before."[2] Such results also should have been no surprise; a public land commission had laid out the basic problem for everyone to see as early as 1880:

> There was a kind of homogeneity in the quality and value of the land of . . . [the Old Northwest Territory]. It was all valuable for agriculture and habitation, but in the western portion of our country it is otherwise. Its most conspicuous characteristic from an economic point of view is its heterogeneity. One region is exclusively valuable for mining, another solely for timber, a third for nothing but pasturage, and a fourth serves no useful purpose whatever. . . . Hence it has come to pass that the homestead and preemption laws are not suited for securing the settlement of more than an insignificant portion of the country.[3]

The final effort to solve the problems of the Homestead Act was the Stock-Raising Homestead Act of 1916, which raised the maximum settlement acreage to 640 acres. But this increase was still wholly inadequate; a later post mortem reported that half of the 50 million acres disposed under the Act were eventually abandoned for farming. Large parts of the lands remaining in farming could barely make it. The range had been plowed up and greatly damaged for future grazing use over wide areas.[4]

In sum, despite much popular glorification of the Homestead Act of 1862, the verdict of historians on it and other public land laws of the next fifty years has been almost uniformly negative: "The Homestead Act of 1862 and its succeeding modifications came too late to fit the area remaining open to homesteading. The result was unnecessary human hardship and failure, deterioration of the range, and accelerated erosion." The public land system generally "encouraged and invited conflict and economic instability, overgrazing, and attempted cultivation of submarginal lands," leading to an "incalculable waste of land and human resources."[5]

Continued Disposal of Federal Coal

The consequences of the federal coal laws have not been nearly so dire, but only because federal coal was not very important until recently. In other respects, however, earlier patterns were closely followed. The coal laws were ignored or evaded and produced results, in practice, almost the opposite of the congressional intent. Despite the provisions for retention and active government management in the Mineral Leasing Act of 1920, federal coal for practical purposes continued to be disposed to the private sector much as it had before the Act. Once in private hands, federal coal was developed according to the forces of the market—with little federal influence asserted.

After passage of the Mineral Leasing Act in 1920, the public lost interest in federal coal and a long period of neglect followed.[6] There was little demand for federal coal after 1920. Moreover, the trend was downward; the biggest source of demand, the railroads, dried up altogether after World War II when coal-burning locomotives were replaced by diesels. Hence, for many years few federal coal leases were sought. In 1922 and 1926 only one federal coal lease was sold; only twenty-three leases in total were issued during the 1920s and twenty-seven leases in total during the 1930s. Up until 1960 the average number of federal coal leases issued was 4.3 per year.

From 1960 to 1969 the number of federal leases shot up to 313, an average of 31 per year. However, most of the leases were still being held for speculative purposes; the majority of the leases ever issued (70 percent) had never produced any coal as of 1973. Where coal had been produced, it was usually in small quantities for local or railroad use, reflecting the absence of any large-scale demand for western coal.

The Interior Department showed little interest in the administration of coal leasing. Under the Mineral Leasing Act a tract could be offered for lease either on a specific request from an applicant or at the initiative of the government. As of 1970 every lease had been offered under the former procedure; there had never been a tract put forth for lease at the specific initiative of the government.

The provisions of the Mineral Leasing Act for diligent development of leases were ignored. Despite the explicit requirement of the act, no lease was ever cancelled for failure to be diligently developed. One problem was that the federal government never got around to issuing regulations defining what diligent development meant. Another complicating factor was that an operator could comply with the provision for maintaining continuous operations by paying an advance royalty. For many leases the required advance royalty was set at $1 per acre, the same amount as the lease rental payment, and any rental payments could be counted against advance royalties. In short, paying the rental meant the provision for continuous operations was satisfied.

The federal government collected little revenue from its leases. There was hardly any bidding competition for federal leases, reflecting the fact that in the typical case only one party was interested in a tract. Between 1920 and 1974, 63 percent of the competitive leases sold had only one bidder. Nine percent sold had no bidder and were simply awarded to the one applicant without any payment. Only 5 percent of the leases sold had four or more bidders. On leases with no more than one bidder, the average bid was just $2.84 per acre. A typical coal acre in the West might well contain 10 to 20 thousand tons of coal and could contain as much as 200,000 tons, implying typical bonus bids of less than 0.03 cents per ton and in some cases possibly as low as 0.0014 cents per ton. From 1954 to 1971 the total bonuses collected from coal leasing were only $13.8 million.

Royalty rates were somewhat irrelevant, given the low rates of production, but royalties were not very high either. The BLM state offices were left to their own devices to administer federal coal leasing and royalty rates in the 1960s varied across states from 10 cents to 22.5 cents per ton. At a typical western coal price of around $4 a ton in the 1960s, royalties ranged from 2.5 to 6 percent of coal value. From 1920 to 1974 the total of all coal royalties collected was only $38.9 million.

There was no competitive bidding for leases issued under the preference right system. One might have expected the number of preference right leases to decline over time, as more information was accumulated about coal geology and there was less need to offer an inducement to explore. But preference right leases instead became a greater share of leases issued, rising from 17 percent in the 1920s to 59 percent in the 1960s, when lease demand was heaviest. Coal companies in the 1960s apparently hit on preference right leasing as an inexpensive way of acquiring federal coal at a time when the prospect of real bidding competition arose for the first time. Previously, it had been a matter of indifference whether a lease was obtained competitively or through preference right leasing, because there would be little bonus payment necessary in either case. Of all the leases issued between 1920 and 1974, 50 percent were preference right leases.

A 1975 study by the Federal Trade Commission characterized federal coal leasing up to 1971. It found that the decision to lease was almost solely the result of industry preference: "All leasing, both preference right and competitive, has been in response to the motions of interested private parties." More specifically, "in addition to letting industry control the timing of Federal lease sales, the Interior Department has also allowed industry to designate the location of each lease." At each site "the sizes of leases have also been set according to industry demands." After obtaining the lease, "the Department of the Interior has thus far largely allowed production from Federal leases to proceed at the lessee's preferred pace." Finally, federal leases had effectively been in perpetuity because, although "diligent development and continuous operation are a requirement of the Mineral Leasing Act of 1920," in practice "the Interior Department has never enforced these requirements." The overall result has been a system in which "over the past 50 years, the Interior Department has in effect allowed industry to control the timing, location, size and type of Federal lease sales." Companies have acquired coal as needed from the federal government and then "production on Federal lands has . . . been allowed to proceed in response to market forces."[7]

As carried out in practice, the Mineral Leasing Act of 1920 actually amounted to a continuation of the disposal policies of the Coal Lands Act of 1873. A few constructive changes were made, however, such as the elimination of the 160-acre limitation which had helped to make the earlier act unworkable. Without severe acreage restrictions, the Mineral Leasing Act now made it possible for coal companies legally to acquire much larger amounts of federal coal, whereas previously they had often felt it necessary to engage in land frauds and other extralegal practices to obtain such coal. The Mineral Leasing Act thus paradoxically served to introduce a freer market in federal coal and to legitimize private acquisition and exchange in this market. This result was accomplished in a law which formally sought to substitute public for private control, was strongly motivated by anti-big-business and anti-monopoly feelings, and, in spirit, really mandated central planning of federal—and, therefore, western—coal development.

The Suspension of Federal Coal Leasing

By the mid-1960s it was becoming apparent to close observers that western coal might well be a much more important source of energy in the future than it had been in the past. The low sulfur content and consequent environmental advantage of western coal might some day make up for its much higher transportation costs to midwestern utilities. In the West good hydroelectric sites were becoming harder to find and environmental opposition stronger. An especially prescient observer might have calculated that oil and gas would eventually prove too valuable to continue to use in power generation, causing a massive switch to coal burning.

The fact that these factors were becoming more apparent to a few people is seen in the rapid increase in the rate of federal coal leasing in the 1960s. Sixty-two percent of all the federal coal leases ever awarded were issued in this decade. In the peak year, 1967, seventy-two leases were sold—more than in most previous decades. Bidding interest also picked up; in Wyoming the average number of bidders after 1960 rose to 2.5. Greater competition was reflected in higher bonuses, which averaged $5.03 per acre between 1960 and 1965 and then jumped to $56.03 per acre between 1966 and 1971.

As a result of the pickup in leasing activity, the total reserves of federal coal under lease climbed rapidly. Indeed, by 1971 the amount of federal coal under lease had reached the huge amount of 17 billion tons—enough to supply all national coal production for twenty-five years. Prospecting permits had also been issued that might ripen into preference right leases containing an additional vast amount of coal, estimated at 9 billion tons. It was obvious that most of these coal holdings could not be brought into production for quite some time and were being held for speculative purposes.

Having paid very little attention to its coal resources for many years, it is not surprising that the federal government was caught largely unawares by these developments. The FTC comments that "it seems clear that Interior did not anticipate the rising value of Federal coal with as much foresight as private companies and individuals."[8] The huge increase in reserves under lease finally caught the attention of the Interior Department in the early 1970s. A study of BLM in 1971 reported that 773,000 acres were under lease at the end of 1970.[9] The study expressed concern that these leases contained huge reserves of coal, yet showed little prospect of rapid development; 91.5 percent were in a non-producing status. The Interior Department had barely begun to focus on appropriate policies for coal leasing. Confronted with the obvious speculative purpose of recent lease acquisitions, the huge reserves in nonproducing leases, and the lack of any firm federal management, the Interior Department informally suspended coal leasing in May 1971. This action set the stage for a new era in federal coal policy.

Signs of a Boom in Western Coal

From 1957 to 1962 western coal production actually fell by 14 percent. The next five years showed a 41 percent increase, but still left western coal production in 1967 only 5 million tons greater than it had been ten years earlier. However, western coal production from 1967 to 1972 more than doubled, reaching 60 million tons. There were a number of reasons to expect a further rapid growth.

The enactment in 1970 of the Clean Air Act created new incentives for significant shipments of coal from the West to the East. Under the Act, national sulfur emission standards for new power plants were set at 1.2 pounds of sulfur dioxide per million BTUs. This standard could not be met by most eastern coal without the installation of expensive scrubbing systems. However, because western coal was typically much lower in sulfur content, it could frequently be burned within the sulfur dioxide standards without any special emission control equipment. For many midwestern and south central utilities the cost of new emission control equipment to burn eastern coal would be greater than the additional transportation cost of bringing in low-sulfur coal from distant western locations.

Another indication of new prospects for western coal development was the active investigation of the economics of mine-mouth power plants in the West. Such plants could transport electricity to distant urban and industrial users by means of extra-high-voltage transmission lines. Concerned about meeting electric power demands that then were believed likely to grow steadily at 7 percent a year, a consortium of electric utilities sponsored the *North Central Power Study*, in cooperation with the Bureau of Reclamation of the Interior Department. The study examined the possibility for new mine-mouth power plants in the West having as much as 50,000 megawatts of total capacity. Released in 1971, the study identified forty-two potential sites where major new power plants could be located in the states of Montana, Wyoming, North Dakota, South Dakota, and Colorado. The study stated in its conclusion that "in order to provide for future electric and other energy needs, the further development of the vast coal fields of the North Central region of the United States is almost a certainty. . . . The abundant availability of resources with the economy of large-scale development may result in attractive economics for the cost of electric energy at the generating plants."[10]

Although the *North Central Power Study* was an early feasibility study, simply the discussion of the generation of such large amounts of power in the West—along with necessary coal mining, probably in excess of 100 million tons per year—stirred considerable public apprehension. Besides the mining itself, the specter was raised of dirtying the air and other parts of the environment in western states to deliver cheap electric power to citizens of states in other regions.

Other studies at the time projected a need for coal gasification plants—which would also use coal mined close by. OPEC's first price shock occurred in 1973–74, creating widespread concern about future U.S. energy supplies and a desire to rely to a much greater degree on domestic energy sources. By 1974, a widely publicized report found that coal development was "going west" with the potential that "the economic and social patterns of the affected states will be changed from rural agrarian to mining, and perhaps to urban industrial."[11]

3. The Strong Opposition to Western Coal Development

The prospect of a major social transformation in the Rocky Mountain West was bound to be unsettling to many people. However, similar transformations have occurred at many times in many places to many people. Despite the personal dislocations experienced, they tended to be accepted in the past as part of the rules of the game, virtually inherent in the nature of social change. Implicitly, it seems likely that most people accepted such dislocation because they believed in social progress and were willing to make some sacrifices. They would play their part in achieving growth and development—also partly because in the long run they expected to benefit significantly from this process.

But major changes in such attitudes became clearly visible in the United States in the early 1970s. A "no-growth" movement emerged which sought openly to insulate particular localities and regions from forces for change. Rather than continuing progress, new prophets instead foresaw a doomsday not so far off in the trends of modern economics and science. This set of beliefs constituted another key ideology—in many ways a religion—that had a major impact on federal coal policy and management.

Each ideology in a way is a lens for observing the world; change the ideology and, much as in changing a lens, the same events or objects take on a greatly transformed appearance. Instead of a symbol of progress and an economic savior for the Rocky Mountain West, the same coal development thus could also appear as the desecration of the purity of nature by the instruments of modern civilization. Seen this way, evil (coal development) threatened to triumph over good (the natural environment), another instance of the Faustian bargain in which modern man appears to many people to be trapped.

The Environmental Gospel

Every movement has its saints and heroes. For modern environmentalism one of the leading heroes has been David Brower, executive director of the Sierra Club for many years. Well before environmental causes became fashionable, Brower led epic battles for the preservation of natural features. In 1956 he blocked the inundation of Dinosaur National Monument by a Bureau of Reclamation dam. He led another long, but ultimately successful, fight against a proposed dam that would have backed water into part of the Grand Canyon —a latter-day "Sistine Chapel" according to Brower. These and other actions catapulted him to the head of the honor role of modern environmentalism.

Brower was the subject of a book by John McPhee, entitled *Encounters with the Archdruid*. In their meetings Brower described his beliefs to McPhee:

> We haven't done *anything* well for a thousand years, except multiply. An oil leak in Bristol Bay, Alaska, will put the red salmon out of action. Oil exploration off the Grand Banks of Newfoundland will lead to leaks that will someday wreck the fisheries there. We're hooked. We're addicted. We're committing grand larceny against our children. Ours is a chain-letter economy, in which we pick up early handsome dividends and our children find their mailboxes empty. We must shoot down the SST. Sonic booms are unsound. Why build the fourth New York jetport? What about the fifth, the sixth, the seventh jetport? We've got to kick this addiction. It won't work on a finite planet. When rampant growth happens in an individual, we call it cancer.[1]

McPhee considered Brower's central message to be obviously religious in nature, a form of environmental gospel:

> To put it mildly, there is something evangelical about Brower. His approach is in some ways analogous to the Reverend Dr. Billy Graham's exhortations to sinners to come forward and be saved now because if you go away without making a decision for Christ coronary thrombosis may level you before you reach the exit. Brower's crusade, like Graham's, began many years ago, and Brower's may have been more effective. The clamorous concern now being expressed about conservation issues and environmental problems is an amplification—a delayed echo—of what Brower and others have been saying for decades. Brower is a visionary. He wants—literally—to save the world. He has been an emotionalist in an age of dangerous reason. He thinks that conservation should be "an ethic and conscience in everything we do, whatever our field of endeavor"—in a word, a religion.[2]

Indeed, there were closer connections to traditional religious themes than is often recognized. The Bible contains many apocalyptic and doomsday visions not so very different from those of modern environmentalism. Virtually contemporaneous with the emergence of environmentalism, there was also a great resurgence of interest within fundamentalist Christianity in Biblical foretellings of future catastrophe. One book, which sold millions of copies in the 1970s, offered the prophetic message that "there were certain signs predicted by the Bible which were to herald man's doomsday. Over the past 20 years, world developments have fulfilled most of the conditions set forth by seers in both the Old and New Testaments."[3] Brower, in many ways, offered a secular version of this message. Brower's version was more palatable to an age in which science has carried much greater weight than traditional religion, especially among the governing elite.

The great environmental battles of the 1950s and 1960s involved places and names like Dinosaur National Monument, Grand Canyon, Storm King, DDT, national wilderness legislation, and Redwood National Park. In the 1970s there were to be many more dams, toxic chemicals, oil drilling rigs, pipelines, timber harvests, and other grave threats to the environment to be resisted. The biggest battle of all, however, may well have involved western coal development. In this case a major expansion of coal mining in the West threatened not just one section of a river, or the side of a single mountain, but much of a whole region, the Rocky Mountain West. Moreover, this region was the last in the United States outside Alaska that was still sparsely populated, still had clean air and water in most places, and still had not been thoroughly taken over by "social progress." The ills of modern civilization that had already beset other parts of the United States appeared about to descend on the West with large-scale coal mining. Environmental organizations therefore rallied to the cause, and a decade-long fierce battle was fought over the expansion of western coal mining.

The Threat to the West

The metaphors of the environmental movement often describe the land and environment as though they are living things, indeed sometimes sacred objects. Development of the land or other man-made changes in the environment are described in terms usually reserved for aggressive, hostile acts by one living thing against another living thing, most often man against man. Thus, the environment is continuously being "killed," "assaulted," "plundered," "murdered," "raided," "stolen," and least of all "threatened." To protect the environment is, therefore, to obey several of the ten commandments, to fight for good against evil. For example, a prominent Montana historian wrote a book to protest and rally the cause against the "rape" of the Northern Great Plains. He suggested that modern civilization proposed to make a "sacrifice" of western lands containing coal to its false god of progress and material abundance. Instead, following the ideas of Aldo Leopold, he proposed a new ethical attitude toward the land:

> When the land and its infinite abundance suddenly (and in a very real sense it *was sudden*, or at least the comprehension of it was sudden) revealed itself to be nearly "used up," the reaction set in. Land, and often land for land's sake, became first a value and then an ethic. That, too is understandable. Because if we have become a great nation by dedicating ourselves to intensive "use," we can hardly avoid a sense of deep loss when we find ourselves running out of much more to use. What do we do now? What, as a matter of fact, are we now?
>
> That is the core of the environmental issue. It is the clash of an old and essentially honored ethic and a new and often strange one.[4]

One of the most heated battles over coal development concerned the construction of two coal-burning power plants at Colstrip, Montana, in the heart of the Powder River region. Speaking at a public hearing, the wife of a local rancher, also a teacher of English, captured the feelings of many residents hoping to make a "last stand" against the project:

> As I drive to work each morning across the Northern Cheyenne Reservation, which is directly in line of flow of the wind from Colstrip, the awesome majesty of the hills feeds my spirit and I feel very close to nature and our neighbors. The scene I see is never the same. The sun, the rain, mist, snow, paint a different picture every morning. But whatever the view, it is always inspiring.
>
> I am not naive to believe that the power company is as humanitarian as it would like us to believe. . . . Montana Power can brush aside a man without a second thought while proclaiming loudly how much they're helping the people of Montana. I have no quarrel with any individual. We are all pawns, to be used by the corporate giant to increase their profits, and when our usefulness is at an end, to be dropped by the wayside. Their stockholders see figures on the annual report, not majestic hills or sandlots. This country took millions of years to form, to create the beauty; and corporate man is going to snatch it away in a snap of his giant shovel jaws. We do not know what the plants will do. We can all make guesses, and the only sure conclusion is that there will be a deterioration in the quality of our life, our water, our air, our land. No one in his wildest predictions sees an improvement in this area. The only disagreement is the degree of destruction.[5]

A second Colstrip resident, a rancher himself, felt much the same feelings, although put in less poetic terms:

> I speak tonight as a minority individual fighting for my way of life. The proposed Colstrip project is an increment with many more to follow in industrialization of a predominantly agricultural region. The social and economic impacts on this region are enormous. Although it is sometimes a matter of personal preference and values, I feel my way of life, the lifestyle of agriculture that we know will be lost forever. My ranch, which can produce crops and livestock and taxes to provide a living, has been doing so for ninety-one years. A power plant has a rather short life. Likewise the coal deposit once mined is depleted forever. Continued taxes and jobs past the life of the plant will be dependent upon other industry or agriculture if there is any left after industrialization.[6]

In other areas westerners on occasion reacted with special bitterness toward the prospect of coal development. One Hopi Indian mourned, "You are taking our water. You are destroying our land. . . . How can we live? It will be the end of our way of life, the end of the Hopi people." A western rancher warned, "Don't make the mistake of lumping us and the land together as 'overburden' and dispense with us as nuisances. . . . Don't be so arrogant as to think you can get away with the murder of the land where tougher and better men have failed."[7]

Many easterners were also much concerned about the impact on the West of massive coal development. One of the most influential studies of federal coal leasing was *Leased and Lost*, published by the New York–based Council on Economic Priorities (CEP) in 1974. The study found that "the environmental quality that has made the intermountain West a recreation capital for the nation will be substantially degraded." One of "the major problems of coal development" will be that "strip mining coal is destructive, and in many places it is irredeemably destructive. . . . The natural ecological system is irreparably destroyed by stripping." Stripped areas in the Southwest perhaps ought to be thought of as "National Sacrifice Areas." The damage there might best be considered permanent since major uncertainties had not been resolved as to "the true reclamation potential of stripped land." Coal mining would have a great impact on the social as well as the physical environment of the West: "Large scale coal development in the West will have a staggering economic, social, cultural, and environmental impact on that region." The study reported that: "It is probable that in the next quarter century, several cities of one hundred thousand people will spring up in the Northern Plains in the wake of massive coal development."[8]

The Strategy of Environmental Opposition

In opposing western coal development, environmentalists could draw upon a broad range of experience all over the United States in halting the construction of dams, apartment houses, highways, pipelines, and many other kinds of projects. By the 1970s a basic set of tried-and-true tactics had evolved.

Prior to the saving of Dinosaur National Monument, the greatest environmental cause had concerned the building of another dam. This dam also threatened to flood a large area of a national park, in this case the Hetch Hetchy Valley of Yosemite National Park.[9] The preservationists of that time were led by a Brower predecessor in the Sierra Club, the club's founder, John Muir. In the earlier case, however, the battle was lost when the dam was approved by Congress in 1913. Muir had been opposed by an equally fervent true believer, Gifford Pinchot. As discussed earlier, Pinchot's faith was the conservationist "gospel of efficiency"; he believed devoutly in the desirability of maximum use of land for human purposes guided by rational scientific calculation.

Brower may well have learned from Muir's failure, because he adopted a much more pragmatic resistance strategy. Rather than engage in a holy war, pitting the values of preservation of nature against efficient human use, Brower decided to contest conservationism on its own terms; he sought to show that proposed dams and other projects failed by the conservationist test of rational efficiency. If he had made it a question of gospels—of maintaining the purity of nature versus rational human use—Brower would in all likelihood have lost. But by actively challenging the scientific basis for proposed projects, he might win, and indeed often did.

Brower's task was made much easier by the historic tactics of Pinchot and other conservationists. Proselytizers themselves as much as scientists, they had made greatly exaggerated—virtually utopian—claims for the benefits of comprehensive planning, the ability to forecast accurately, and generally the level of scientific rationality that could be achieved in public (or private) management. Indeed, all Brower needed to do in many cases was to contrast the thin scientific justifications and uncertain rationales actually offered for most dams and other proposed projects with the very high standards for rational planning of conservationism. In short, when Brower suddenly asked whether the emperor had any clothes, it often proved exceedingly difficult to find them.

Such tactics were formulated in the battle over Dinosaur National Monument and perfected over the next decade. In the battle Brower, who had confessed to McPhee that he regarded figures as mostly a matter of the right "feeling," nevertheless was more than willing to use "facts" and "science" in the service of the cause:

> The Dinosaur Battle is noted as the first time that all the scattered interests of modern conservation—sportsmen, ecologists, wilderness preservers, park advocates, and so forth—were drawn together in a common cause. Brower, more than anyone else, drew them together, fashioning the coalition, assembling witnesses. With a passing wave at the aesthetic argument, he went after the Bureau of Reclamation with facts and figures. He challenged the word of its engineers and geologists that the damsite was a sound one, he suggested that cliffs would dissolve and there would be a tremendous and cataclysmic dam failure there, and he went after the basic mathematics underlying the Bureau's proposals and uncovered embarrassing errors. All this was accompanied by flanking movements of intense publicity—paid advertisements, a film, a book—envisioning a National Monument of great scenic, scientific, and cultural value being covered with water.[10]

By the 1970s the tactics of the Dinosaur battle were well honed. As a result, the opposition to western coal development was not put primarily in moral or preservationist terms. Rather, a blizzard of arguments spewed forth from the early 1970s on questioning the true scientific rationality of western coal development. As in many other cases, the government found it very hard to respond, to demonstrate in any objective way that the social benefits would really exceed the social costs.

The Scientific Side of Environmentalism

Muir is often included with Pinchot as part of the early conservation movement, despite the fact that, after an initial friendship, they came to be bitter enemies. The modern environmental movement has had not only its Browers but, also, its heirs to the Pinchot tradition of conservationism described above in chapter 1. There have thus been many environmentalists whose greatest commitment has been a rededication to technical expertise for maximum human use, to comprehensive planning, and to the more truly scientific conduct of government affairs. One major concern of such people has been the irrational neglect of environmental benefits and costs in a great deal of government policy making. A common government error was to consider only marketable outputs and privately borne costs in its calculations. A new government decision-making calculus hence was urgently needed which would incorporate the real value of a clear sunset, of clean water, or of a lushly wooded forest, as well as the real costs of environmental damages done by chemicals and other pollutants.

The two camps of the modern environmental movement have managed to maintain an often productive alliance against their many common enemies —ranchers, miners, timber companies, oil companies, and other commercial "special interests." The ability to sustain the alliance has been greatly aided by the particular tactic of the Brower camp to employ scientific rationality to attack undesirable projects—however tenuous this camp's actual commitment to such rationality. The modern heirs to the scientific management precepts of Pinchot conservationism have in fact often labored as energetically as Brower —and brought much greater technical skill and expertise to the cause—to expose the technical failings of public agency justifications offered for proposed projects.

Indeed, the scientific rationalists have also been morally outraged on many occasions—in their case by the many technical deficiencies of real world project justifications matched against the ideals of true science. The great enemy of conservationism was politics, and the unwarranted intervention by special interests in scientific decision making by experts. For Pinchot the purity of public decisions often seemed to take the place of the purity of the woods for Muir. Not only have many contemporary project justifications been scientifically deficient, but it has also been clear that there was never any real intent to make them so. Rather, their real purpose has been to provide a thin veneer of rationality for decisions made for political reasons at the behest of special interests.

In short, not only could the two contemporary camps of environmentalism join together in a common undertaking to expose scientific and technical inadequacies, but in many instances they could do so holding hands in a joint spirit of moral outrage—even if for much different reasons. The only problem would be if a project came along which was justified in a suitably rational and scientific manner. Perhaps there would be some such projects some day. But the problem did not arise very often. Moreover, it was easy to avoid a divisive conflict, because the standard for scientific "adequacy" was inevitably very subjective and tended to be set very high. It was always possible to call for more analysis, more data, more scientific investigation, better forecasts, etc. The scientific approach was itself not applied very often to the question of whether more scientific analysis is always justified, in short, whether the benefits of more analysis will actually exceed the costs.

It is also likely that scientific environmentalism provided a respectable outlet for some people who really were motivated by less rational concerns. Such people perhaps were really drawn to the Brower gospel, but were also steeped in traditions of rational science; it was too much in one fell swoop to jump from a strong commitment to hardheaded scientific rationalism to a rejection of scientific progress, industrial and urban development, and many of the basic tenets of modern society. An acceptable compromise thus was in practice to attack most new projects and other applications of modern technology, but to do so in the less threatening name of science and rational analysis itself.

The Scientific Challenge to Western Coal Development

Following the pattern, by then well established, the environmentalist opposition to western coal development took a highly scientific form. The most important and influential such challenge was mounted in a report prepared by the Institute of Ecology in the form of a comprehensive set of public comments on a draft environmental impact statement (EIS) released in May 1974 for a new federal coal leasing program. The comments were edited by Katherine Fletcher, then on the staff of the Environmental Defense Fund, and a few years later to become a staff member on the White House Domestic Council in the Carter administration.[11]

The EIS comments recognized that federal coal ownership created an unusual situation with the potential for planning a major new industry, western coal production. The EIS commentators showed few doubts about how this planning should be done; the government had an obligation to prepare comprehensive formal plans to guide directly a future rational pattern of western coal development. Indeed, it would be a rare chance to put the longstanding conservationist precepts of comprehensive land-use planning into practice at a national level: "This places the nation in a unique position to plan coal development through its elected representatives, bringing social, cultural and environmental impacts of development under consideration to supplement the traditional short-term economic viewpoint of private industry. The coal leasing program could be the model for implementation of the first national land use and energy strategy." Under this strategy, "rather than a coal *leasing* policy, the government needs a policy to control coal development."[12]

The commentators on the Interior coal-leasing EIS were not lacking in suggestions about the proper direction of government influences on western coal development. They offered a long list of reasons why western coal development should be deemphasized. Recognizing that market forces were clearly pushing coal production westward, the commentators sought to raise a number of possible reasons why the market might be wrong: "The [Interior Department] EIS seems to accept the 'conventional' wisdom of the companies involved in western coal development, and fails to consider two factors of utmost significance: (1) the many disadvantages of western coal development; (2) the available alternatives both of low sulfur coal in the East and of sulfur control technology. Taken together, these factors constitute a serious challenge to the course of all-out western coal development." Furthermore, the comments suggested that western coal is less economical than often thought because it has a much lower heat content than eastern coal. High transportation costs were subject to automatic adjustment in utility power rates and thus "increasing reliance on western coal appears not to be based on economic efficiencies." The advantages of low-sulfur coal in the West were not as great as widely believed because there was strong competition from "very large quantities of low sulfur coal [which] exist in the eastern United States." Recognizing that western surface mines were much safer for workers than eastern underground mines, the study ventured that nevertheless there was a major untapped "potential for safety improvement" in the East. Finally, "movement of the industry from East to West not only creates a boom where it goes, but also leaves unemployment in its wake. And, of course, the substitution of capital and energy intensive strip mining for labor intensive deep mining has a negative affect on overall employment."[13]

The EIS comments did not actually estimate the cost of eastern versus western coal mining for various power plant locations, environmental constraints, and other circumstances. Confined to qualitative observations, the comments considered that the burden of making detailed cost and other estimates fell on the government. The government had to accept the obligation to demonstrate that western and federal coal was in fact superior to eastern coal for the markets it might serve. The draft of the coal-leasing programmatic EIS proposed by the Interior Department had not refuted the many objections raised to the desirability of western coal development; rather, its "authors appear to advocate blanket encouragement of large scale western coal development, in the complete absence of the appropriate information and analysis." As a result, "not only does renewed federal coal leasing need to be reconsidered, but also the wisdom of developing many existing leases must be closely questioned."[14]

The EIS comments by the Institute of Ecology criticized the Interior Department's leasing program because "as it presently stands, it appears it will have to rely primarily on the planning of the private sector. This would be inappropriate, not necessarily because of defects in private planning, but because the government, as sovereign, has obligations which transcend the traditional obligations of the private energy industry." Indeed, the federal government was obligated to plan the full national energy sector. Otherwise, how could the necessary part to be played by federal coal be determined? As the EIS comments put it, various "energy demand alternatives must be systematically described, perhaps as alternative 'scenarios.' The role of coal in each of the possible demand configurations should be described." Only after such a comprehensive examination of national energy options would the production of federal coal come up: "*Then*, the role of federal coal—existing leases and/or new leases—should be examined, making clear what the choices and assumptions are. It is only in this context that the program alternatives open to the Department can be meaningfully considered."[15]

The writers of the EIS comments finally also came to an unsurprising bottom line; they concluded that there would be little need for new federal coal leasing for many years: "The fact that more than 50 billion tons of federal and non-federal coal reserves have been committed in the Northern Plains area alone indicates that quantitatively far more western coal is presently under lease than would be required to meet all foreseeable needs for at least the next few decades. An obvious conclusion is that renewed federal leasing cannot be justified on the basis of need for federal coal to meet U.S. energy needs." The Interior Department was challenged to disprove this contention: "A major object of the impact statement should be to consider the *need* to institute a leasing program, if indeed leasing is proposed. . . . Unfortunately, the EIS fails . . . to demonstrate the need for renewed federal coal leasing." The standard for moving forward with leasing was not simply a reasonable expectation of need; rather "in the absence of . . . compelling proof, the rationale for the Department of the Interior proposal to renew leasing collapses."[16]

Besides a clear demonstration of an actual need for both western coal development and new federal leasing, the Interior Department should also have a fully developed program for carrying out its leasing prior to the resumption of any further leasing. The EIS comments found that it was "readily apparent that the Interior Department has nothing which can be described as a tentative coal leasing program." The comments concluded that a brand new programmatic EIS should be prepared and the existing moratorium on federal coal leasing should be continued until the following six conditions had been met:

1. the need for further leasing is established;
2. strong federal laws exist regulating surface mining and protecting private surface owners;
3. existing leases and preference right lease applications come under sound environmental and management practices;
4. mineral leasing laws are revised to ensure environmental quality and protection of the public trust;
5. Project Independence has been fully debated and evaluated by the government *and* the public; and
6. basic information on coal and coal development impacts is both understood and evaluated.[17]

The EIS comments prepared by the Institute of Ecology were only the most prominent and the most sophisticated of many such environmental critiques. The study of federal coal leasing noted above by the Council on Economic Priorities had similarly argued that further western coal development should await an adequate set of institutions for planning and controlling it. The CEP study suggested that since "development is proceeding in a random, uncoordinated way," and "coordination and regional planning are virtually non-existent," then "in view of the vast supplies of coal in the East and Midwest, it is doubtful that haphazard western coal development is necessary or wise."[18]

The CEP study also recognized that strong market forces were propelling western coal development forward, that "the coal industry is moving west." Moreover, federal coal would be central to the shift: "The public and Indian leasing programs are the keystone of a coal development boom that is predicted for the West." The Interior Department had thus far entirely failed to examine whether it should try to hold these forces back or how to control them better: "Interior has refused to plan for or even to consider the environmental, social, cultural, or economic effects of its leasing program." A whole new approach to leasing thus was required based on the recognition that "national resource use planning is needed" to direct western coal development.[19]

An important step would be to establish control over the development of previously owned federal coal. The CEP study considered that past federal coal leasing—"an unplanned, massive, permanent transfer of public resources into private hands"—had already "saddled the nation with a huge block of leased but unmined coal that may well frustrate energy resource planning for decades to come." Rather than through government planning, federal coal already under lease would now be developed at the "whim of industry."[20]

Environmental opponents of federal coal leasing in essence were demanding nothing more than that any future program must live up to the conservationist standard of scientific rationality. The Interior Department could hardly ignore such criticisms because in fact the laws and other existing institutions of the public lands were themselves founded in conservationism. Moreover, much of the wider public considered that government both had an obligation to and could in fact actually demonstrate in a clearly objective and rational way the reasons for its policies. By 1974 the Department was actively examining the question of how to respond to widespread unhappiness with past leasing practices and the likelihood of strenuous environmentalist opposition to any future renewal of federal coal leasing.

Part II. A Coal Program Based on a Planned Market

Introduction

The May 1971 informal moratorium on federal coal leasing had been imposed by the Interior Department in order to provide breathing space to design a permanent leasing program. The Department moved very slowly, however, and little happened until early 1973. From then on coal-leasing matters finally began to receive substantial attention from high-level officials in the Interior Department. Much of this attention was directed at the specific design of a program for leasing federal coal.

Top Department officials recognized the conservationist foundations contained in the Mineral Leasing Act of 1920 and the policy implications for federal coal policy. However, they considered the policy prescriptions of conservationism to be in many respects misguided; in particular, they saw the central planning prescribed by conservationism as inferior to the market mechanism as a means of allocating resources. Since conservationism was written into law, however, and there were few political prospects for changing the law, it would obviously be impossible to ignore. The basic strategy became to make the minimum concessions to conservationist ideology legally necessary, while seeking to design a coal-leasing program based on a sounder concept. The foundations for the new coal program were provided by what I call in this book the ideology of "planned-market liberalism." This ideology emerged after the central planning approach of conservationism and an old-style, free-market approach were both rejected as inadequate or infeasible.

4. The Rejection of Central Planning and the Free Market

In February 1973 the Secretary of the Interior, Rogers Morton, announced that a brand new program for leasing federal coal would be developed. The agency responsible for managing federal coal, the Bureau of Land Management (BLM), would undertake this task. The BLM had been formed in 1946 by a merger of the Grazing Service and the old General Land Office—the latter agency by then with little to do, public land disposal having ended. The central concern of the newly formed BLM was the management of the public rangelands under the provisions of the Taylor Grazing Act of 1934. Until the 1970s the main efforts of the BLM had been devoted to improving rangeland conditions and promoting better grazing practices among western ranchers who used the public rangelands.[1]

The BLM has never achieved the wide influence and public respect of the Forest Service. Nevertheless, the main principles of BLM land management are essentially derived from the Forest Service, and can be traced back to Gifford Pinchot and the conservation movement. However formidable the obstacles have often been in practice, the guiding ideology of the BLM has been scientific management to promote rationally efficient grazing and other use of the public rangelands. The BLM, like the Forest Service, has believed in the effectiveness of forecasting and planning and in recent years both agencies have devoted large funds and placed a great emphasis on formal land-use planning.

The BLM, in its early years, was an agency often short on resources, receiving much less budget and personnel than the Forest Service, even though similar amounts of land are managed. Rancher users of the public rangelands had held BLM influence to a minimum by keeping the agency short on funding. This was one of the main reasons why the BLM had paid almost no attention to federal coal management until the 1970s, despite the signs in the 1960s of its growing importance. Hence, in 1973 when the BLM was assigned responsibility to design a new program for federal coal leasing, it had neither many people nor much experience in matters other than livestock grazing. The program it designed was described in only the barest terms. It was, in fact, a very easy target for the environmentalist torrent of technical and scientific criticism that was about to descend on federal coal-leasing efforts.

The EMARS Coal Program

Nevertheless, in its very brief sketch of a program, the BLM offered a reasonably faithful application of conservationist principles to the specific problem of federal coal leasing. True to the principles of scientific management, the heart of the BLM proposal was to undertake extensive planning, both of energy and land-use matters. The new federal leasing program was called the Energy Minerals Allocation Recommendation System (EMARS), and was described in a draft programmatic EIS released in May 1974. The three main components of the program were "allocation," "tract selection," and "leasing." The program employed an "allocation process to determine the rate at which inventoried federal coal should enter the market." But only the following very brief explanation was provided:

> The *allocation process* relates inventoried federal coal resources to projections of coal-derived energy needs. Total national energy needs are disaggregated into regional demands for coal-derived BTU's. These data, along with any policy directives as to the overall role of federal coal in the total energy mix, are obtained from various Departmental agencies and fed into the empirical federal coal allocation model. The system allocates regional demands for federal coal resources to specific inventoried coal resource areas. The allocation to federal coal resource areas will be on the basis of the least cost of delivered energy and takes into account identified known supplies that have not yet been committed.[2]

Tract selection was to "relate these demands to optimum sites where the best coal can be equated with the most favorable rehabilitation potential." Again, the total explanation offered consisted basically of the following one paragraph:

> In the tract selection phase, the coal allocation targets will be distributed to coal-leasing States and Districts. . . . At the District level, Bureau of Land Management minerals personnel, coordinating with the Geological Survey, will "lay out" optimum coal lease sales containing the targeted amount of high-quality reserves, in areas where effective rehabilitation can be assured, and then prepare Mineral Activity Plans according to established procedures of the Bureau of Land Management planning system. After public participation, a final planning system multiple-use decision will be made by the Bureau of Land Management's District Manager, which will include a site-specific allocation recommendation (AR). This will be coordinated with other Districts' submissions at the State level. Each (AR) will include definite rehabilitation objectives chosen from the alternatives available, and financed at optimum levels from coal production. Base resource data will be adequate in all cases. Allocation recommendations (AR's) from the States will then be aggregated at Bureau of Land Management Headquarters level (Washington,

D.C.) into a site-specific, 1-year leasing schedule and a tentative 4-year leasing schedule, and then submitted to the Secretary for final consideration, adjustment and approval.[3]

The actual leasing process was described in a similarly abbreviated fashion. Probably hoping to make up for the lack of any program details, the bulk of the EIS then went on to present an elaborate description of mining techniques, vegetative types, socioeconomic circumstances, and other matters, but it contained scarcely any policy guidance.

Not surprisingly, the response to the draft EIS was uniformly critical, some of which was discussed in the previous chapter. Even within the Interior Department, the Assistant Secretary for Program Development and Budget commented, "I have reluctantly released the draft EIS on the coal program. In my opinion, it still has major weaknesses."[4] The Environmental Protection Agency recommended starting all over with a new draft EIS because "the EIS barely touches the question of BLM management alternatives and the tradeoffs of environmental impact that might be expected."[5] The chief industry spokesman, the National Coal Association, was not much more complimentary: "The EIS is extremely general and many of the essential aspects and details of the proposed Federal Coal Leasing Program are not set out with any degree of specificity. As a result, it is extremely difficult to completely understand precisely what EMARS . . . is or how it will function, or its relationship to the Bureau of Land Management's Land Use Planning System."[6]

Comments from various environmental organizations were, of course, critical. Indeed, by their standards of comprehensive planning and scientific rationality, the programmatic EIS which BLM had released was so bad that it could only be considered laughable.

The Interior Rejection of a Conservationist Approach

In order to overcome wide opposition, by the spring of 1974 top Interior Department officials had concluded that the description and explanation of the federal leasing program would obviously have to be drastically improved. It appeared doubtful that BLM could accomplish this task, certainly not without considerable assistance. Moreover, there was little confidence among Department officials in the basic wisdom of the type of coal program that BLM had proposed.

Hoping to exercise tighter supervision of the various Interior agencies, the Interior Department in 1973 had upgraded its analytical capabilities at the secretarial level. A new Assistant Secretary for Program Development and Budget was created, including an Office of Policy Analysis. A second new Office of Mineral Policy Development was placed under the Assistant Secretary for Energy and Minerals. These two new policy offices were heavily staffed with economists and public policy school graduates new to the Interior Department. The Assistant Secretary for Program Development and Budget was given a leading role in the redesign of the coal program, and the Office of Policy Analysis provided much of the staff support. The major influence of the Assistant Secretary for Program Development and Budget partly reflected the fact that the holder of this office, Royston Hughes, had long been an aid and close associate of the then Secretary of the Interior, Rogers Morton.

The assignment of economists to redo the federal coal program was bound to have significant consequences. Where the BLM philosophy was based on the conservationist tradition, the Interior economists had been trained in economics graduate schools; rather than Gifford Pinchot, their ideas traced back through John Maynard Keynes, Alfred Marshall, John Stuart Mill, David Ricardo, and finally to Adam Smith. By long tradition a public land manager is trained to think of scientific management achieved through resource projections and various other formal planning procedures. By a still longer tradition an economist is trained to think of resource allocation through the market mechanism. The emphasis in economics is on competing social needs and the resulting conflicts over the allocation of resources; economists see the basic economic problem as one of determining the appropriate mechanisms for resolving such inherent conflicts among claimants on the social product. Conservationists, in contrast, tended to ignore such conflicts, as reflected in Pinchot's famous dictum that the Forest Service goal was to achieve "the greatest good of the greatest number in the long run."[7] To an economist this looked like a meaningless platitude which attempted to avoid the reality of competing claims on resources, both among different individuals, interest groups, and between the present and the future. Harvard Professor Edward Mason, a leading economist, once stated that conservationism "was a political movement with objectives as disparate as saving the forests, destroying the monopolies, and maintaining Anglo-Saxon supremacy. Its economic analysis was practically nonexistent, although it did emphasize the importance of sustained yield in renewable resources. The best it could do in defining the meaning of conservation was to say that it meant a "'wise use of resources.'" But "almost any program could be accommodated under this rubric."[8]

In recent years economists have leveled numerous severe criticisms at the economic irrationality of conservationist principles, especially as carried out in Forest Service land management. Marion Clawson of Resources for the Future, one of the best known U.S. resource economists, has called the Forest Service management of the national forests economically "disastrous."[9] He contended that "the national forests emerge as a great feudal estate, land poor, managed extensively, relatively unproductive. It is difficult to visualize any values resulting from the present capital intensive forest management that would not be available in equal or greater amount in a forest management program which economized on capital use."[10] Another well-known economist, now at the Brookings Institution, Anthony Downs, wrote that "societies often find themselves in cul de sacs where progress is possible only if irrational beliefs are discarded. In my opinion, sustained yield forest management provides an excellent example of this situation."[11] Leading environmental economists, as well, generally shared such views. John Krutilla considered that while foresters have "appreciated that there is need to provide the 'correct' level and mix of the various resource services which the National Forests are capable of producing, up until the present the instincts and proper impulses of the profession express themselves somewhat more as high motives and sincere exhortations than as the application of operational criteria directed to the optimal mix of wild forestland outputs."[12]

Economists thus had long been very critical of conservationism and its lack of any economic underpinnings. At heart it seemed more an ethical or religious movement with its own gospel than a serious study of economic questions. Rather than careful calculation of benefits and costs, it was directed at defeating the special interest "pillagers" and "looters" of "the public's" natural resources by instituting government planning to achieve a utopian "scientific" resolution of all use conflicts.

It was therefore no surprise that, when the newly arrived economists at the Interior Department encountered the BLM conservationist proposal for extensive government planning of federal coal development, they reacted with great skepticism. It seemed mostly a new example of the old and for them discredited conservationist faith in planning. This impression was reinforced by the very thin sketch which BLM provided for its proposal, which was typical of earlier instances of much greater conservationist devotion to the lofty ideal of planning than to the real world details of its implementation.

The original BLM proposal had included the loosely described procedure mentioned above for allocating national energy demands into regional coal demand and then into requirements for federal coal leasing. As the reassessment of the coal program got underway, BLM sought to flesh this concept out further, explaining how it might go about calculating leasing targets:

> Secure demand curves from the AS/E&M [Assistant Secretary, Energy and Minerals] for each of the coal States in terms of market sectors (i.e., local and export [out of State], electrical generation, coal gasification, and coal liquifaction). In developing the targets for each State (in recoverable tons of coal), consider and equate the following factors: (1) demand curves for each State; (2) how much coal should come from Federal lands—determine ratio of Federal coal to private coal by comparing acres of each in each State or coal producing area for each State; (3) review industry areas of interest and nominations; (4) use appropriate lag factors for each market sector (i.e., 5 years for local and export, 8 years for electrical generation, 10 years for coal gasification and more than 10 years for coal liquifaction); (5) use a 60 to 1 reserve to production rate in terms of recoverable coal; and (6) consider existing coal leases and the need to lease additional coal.[13]

Such a top-down allocation approach required that government estimate national energy demands, disaggregate them into coal demands (total and regional), then determine how much of the coal should be federal and how much nonfederal and, finally, prescribe how much of the coal should come from existing federal leases and how much required new federal leasing. In following these steps the amount of information centrally required would be very great. For example, in deciding the coal production shares of federal and nonfederal coal within a given region, all potential coal tracts should be examined and those with the lowest costs selected, some containing federal and some nonfederal coal (and some both). In determining how much total coal should come from any given region, all the possible customers for regional coal would have to be surveyed and an assessment made as to whether the cost to them of coal from this region was lower than from any other region. Differences in coal quality among regions would also have to be taken into account. The sulfur content, transportation costs, mining costs, and other features of the coal from every region would have to be compared. The question of whether coal from a given region could compete with oil and gas, nuclear, hydroelectric, or other energy sources also had to be answered for each possible destination for regional coal. Total energy demands themselves were important. For example, what would be the level of future electric power consumption? Finally, any major changes made at one part of the energy system would set off reverberations necessitating recalculations throughout the rest of the complex overall system.

To say the least, in 1974 the Interior Department was wholly unprepared to make such calculations. The BLM certainly had no idea how to do them and, indeed, probably did not even fully recognize their necessity under its proposed approach. Even in the best of circumstances, there would remain major questions as to the ability of government to assemble such large amounts of information and to perform the complex coordinating act required to balance supplies and demands simultaneously all the way from the national to the local level.

The Director of the Interior Office of Policy Analysis, Harvard-trained economist William Moffat, sought to head off any such centrally planned system for federal coal leasing. He was convinced that comprehensive advance planning could not successfully be undertaken. The information requirements would be too great, the inherent uncertainties too many, and the computational difficulties too severe. An attempt at fully comprehensive planning in his view would only lead to pretense:

> Some have urged that we make the comparison now, once and for all, and decide which deposits of federal coal should and should not be leased; they argue that if we do not do so, we are not truly "planning" for federal coal leasing, we are only setting up a process without specific content.
>
> I strongly disagree. No such set of decisions made now would have validity years later when the actual question of leasing might arise. To publish such a "plan" would be a delusion, with the threat that it would generate commitments, both for and against development, which could not later be reversed,

and which would do great injury to the public interest. I believe that proper policy involves instead a commitment to make careful, case-by-case comparisons of value and cost, taking account of all information and circumstances that exist at the time. This is what our new planning process is intended to do.[14]

Rather than a comprehensive plan, the need was for a decision-making procedure and sets of standards for case-by-case resolution of questions. This approach would allow information and analytical resources to be brought to bear where and when they were most needed, thus avoiding a basic pitfall in comprehensive planning of trying to solve in advance many problems which actually will never arise—and the corollary that it may therefore fail to solve any because analytical resources are stretched so thin. Under comprehensive planning attention is deflected away from the problems which have already arisen and of which there is no doubt of their true importance.

The staffs of the new Office of Policy Analysis and Office of Mineral Policy Development thus combined their weight to oppose a comprehensive planning approach that would have been true to conservationism. As economists they were instead disposed to follow the market. A market orientation also matched the views of the high-level Nixon and Ford administration appointees in the Interior Department at the time. The freshly redesigned coal program therefore would be based on market principles.

The shift to a market system was symbolized by the prominence now given to coal industry nominations. Leasing efforts would be centered around requests to coal companies to indicate which specific tracts they were most interested in mining. Interior Department environmental studies, geologic examinations, fair market value estimates, and other leasing activities would begin with these tracts. As one Department proponent stated, the new program would be "driven *primarily* by expressions of industry interest," unlike the old program of "demand allocation."[15]

A Critical Obstacle to a Free Market

Gunnar Myrdal has emphasized that the demise of the free market was never really planned; it actually occurred through a series of ad hoc interventions by government combined with pressures within the market towards concentration. Government would discover a practical need to take action in one area, a pressing social concern would force government attention to another matter, special interest groups would successfully push for government assistance in a third area, and so forth. Thus, as Myrdal describes the process, "while admittedly the State and the citizen, step by step, have been taking over more and more responsibility for the direction and control of economic life, they have been led by events, not by conscious choice."[16]

Sometimes the growth in government intervention has even been carried out under the formal banner of the free market. As Galbraith has often pointed out, there has been a strong reluctance in the United States to be explicit about the extent to which the free market has already been superceded by a planned internal resource allocation of the large private corporation—in Alfred Chandler's terms, the extent to which the "visible hand" of corporate decision making has actually replaced Adam Smith's "invisible hand."[17] Similarly, American farmers often talk about a "free market" in agriculture, despite a long history of government price and output controls, as well as numerous other interventions. Indeed, the unusually large government role in agriculture is probably due mostly to the necessity for the government, itself, to perform planning and coordinating tasks that in other sectors are accomplished internally by corporate planners. In short, the stated commitment to develop a market system for federal coal management was no guarantee that a free-market result would really be achieved.[18]

A true free-market solution for federal coal could be achieved by disposing of large amounts of coal to industry for whatever price it would bring at the present time. With much more than enough such coal available to meet near-term demands, industry could then select from the large supplies and in the process decide when, where, and how much coal to produce. As described above in chapter 2, this approach was followed under the early coal disposal laws and then under the Mineral Leasing Act of 1920, as it was actually administered up to 1971, despite the contrary congressional intent.

The standard economic analysis of exhaustible resources emphasizes the point that the free market will efficiently distribute the production of resources from the present into the future.[19] The private owner of a mineral deposit that will be more valuable in the future has the incentive to hold it out of production —i.e., to hold it speculatively—until the point of maximum future value is reached (properly discounted for the wait to receive revenue). Such speculation is indeed considered by economists to be socially beneficial. As one report on federal coal leasing stated, under competitive circumstances "if speculation did occur, it would necessarily reflect that the reservation of this resource for future uses was the most socially desirable current employment pattern."[20] In essence, speculation is conservation, achieved, however, by private profit incentives, in contrast to the altruistic motives for conservation that the conservation movement traditionally espoused. The traditional conservationist has considered that an element of painful sacrifice would inevitably and appropriately be present in giving up current gratification for the benefit of the future. The idea that future users should pay the present to defer resource use—i.e., that prices of resources should rise rapidly enough to induce current owners to keep them out of production—was alien to the moralistic spirit of conservationism.

The modern age prides itself on its scientific rationality and its immunity from old religious and other superstitions. But the attitude of the conservation movement towards speculation finds close parallels in the opposition to payment of interest on ethical and religious grounds in the Middle Ages. Interest on money in the bank performs a function essentially equivalent to speculative gains in resource value; collectively, society provides for its future capital stock by offering interest in exchange for giving up current consumption. When the investment is in the form of a conserved natural resource, the interest payment effectively takes the special form of a gain in capital value. As with other investments, the interest must be greater as the risk increases, i.e., the expected capital gain must be greater the more distant and less certain the future rise in resource value. Moral offense at the idea of interest—money paid just for waiting and bearing risk without any real production—not only aroused the ire of the Roman Catholic Church in the Middle Ages, but even now lies behind usury laws on the books of many states. It also shows up in the Marxist idea that capitalists "exploit" the workers by taking any surplus above the direct labor input into production.

Speculation in federal leases stimulated the same kind of moral offense. One Interior Department official observed that "the current outcries about the Department's past coal leasing policies apparently result from the fact that there are 16 billion tons of recoverable Federal coal under lease while production from Federal leases was only 13 million tons last year."[21] The influential report, *Leased and Lost*, castigated the Interior Department for following a policy described as "public lands and private profit," where the profit was the result of successful past speculation.[22]

As a product of the conservation movement, the Mineral Leasing Act reflected its antagonistic attitude towards speculation. Trying to make speculative holding of leases impossible, the act required the "diligent development" of federal leases. It thereby shifted the responsibility for deciding when federal coal production should start from the market to the government. The point to begin production is one of the most critical decisions for any producer of a coal deposit. Hence, once the diligent development requirement of the Mineral Leasing Act was actually enforced, a true free-market solution to the production of federal coal would no longer be possible.

The Interior Department thus had to consider whether it should ask Congress to amend the Mineral Leasing Act to abolish the diligent development requirement. Such a request would be consistent with the market orientation of the economists now developing a new leasing program. In addition, however, there were a number of other practical consequences of diligence requirements to be assessed, not all of them undesirable.

The Practical Consequences of Diligent Development Requirements

One potential role for diligent development requirements is as a tool of antitrust policy. In the past there had been much concern about the possible accumulation of most federal leases in the hands of a few coal companies, leading to excessive concentration of coal holdings and eventually of coal production. A diligent development requirement precludes this possibility. This element was an important consideration in the original inclusion of such a requirement in the Mineral Leasing Act.

Diligent development requirements also make it easier for each generation to choose a pattern of federal lease development that best satisfies its own environmental and other nonmarket values. If federal leases issued now were still being developed many years in the future, there would be no assurance that they would be the preferred leases according to the values of these later periods. A number of federal leases issued in the 1960s and earlier decades probably would not be issued now, due to changed perceptions of environmental risk and value. According to one Interior Department analysis, "changes in coal leasing policy, especially those connected with environmental protection, may be harder to apply on existing leases than on future leases, so that the greater the leased reserves, the smaller the federal domain over which leasing policy changes can be effectively applied."[23]

The government may also maximize its financial returns by holding onto federal coal until this coal is ready to be developed. Because of its huge size, the federal government is probably much less risk averse than private speculators and hence will discount less the value of any future expected revenue from coal. If the government values future revenues more highly, its coal properties will be worth more to it (financially) than the price they would bring from private speculators. The government, in effect, prefers to engage in speculative holding of its own coal properties, because it is the one speculator to whom these properties are worth the most. This situation seems to have occurred in federal coal leasing. As an official document of the Interior Department said, "The problem has been that in the mid-1960s too great a quantity of Federal reserves was leased, in relation to production, so that too much of the inventory profits from rising coal prices have accrued to lessees rather than to the Federal Treasury."[24] To be more precise, the problem was not the existence of any inventory profit, but that from the federal viewpoint the private rate of profit was too high for the risk and length of investment involved in holding the federal coal, so that the federal government instead preferred to retain the coal.

In judging the wisdom of past federal leasing, however, the public probably tends to misinterpret and overestimate the long-term gains to speculation. The public usually sees only press and other reports of successful speculation and then takes these success stories to be the run-of-the-mill return to speculation. If someone offers a speculative opportunity with a one in ten chance of reaping a 100-percent capital gain, and a nine in ten chance of no return at all, the expected or average return is only 10 percent. However, if the public hears about only the one success, it is likely to conclude that a much higher average return is obtained. Purchasers of federal coal leases in the 1960s may simply have been among the luckiest winners—thanks largely to OPEC price rises, which no one could have foreseen with any certainty.

Countering the positive features, diligent development requirements have some very practical drawbacks. Holders of valuable coal leases could be forced by a diligence requirement to begin production ahead of actual market demands, conceivably by stockpiling production if no other outlet can be found. More likely, coal companies would offer bargain prices simply to retain the lease, departing from the economically most efficient pattern of coal production to start certain leases up early. The inefficiencies that might result from diligent development requirements were suggested in complaints from the coal industry and others to the Interior Department:

> Another unintended but quite likely result of the rigid application of the diligent and continuous operations requirements . . . could be unnecessary environmental degradation.
>
> This is not simply a theoretical possibility. If Western Fuels, for example, must undertake immediately to develop or lose each of the non-contiguous coal leases in which it has an interest—leases which are the result of the acquisition at considerable cost of preference right lease applications in order to provide assured coal reserves—or face their loss, it will obviously develop them. The same will be true of others. As a result, the Department of the Interior, contrary to sound environmental concepts, will have forced the opening of mines which, on a more logical basis, would be postponed until production therefrom is indicated on the basis of carefully coordinated needs.[25]

A diligent development requirement can cause serious complications where there are fragmented coal-ownership patterns. In these areas an often slow assembly of coal development rights must be undertaken, similar in some respects to central city assembly of office building sites. Hence, it may be unrealistic to expect rapid development of federal leases that constitute only one portion of a potential mine. Noting such problems, not only industry but the Environmental Law Institute as well expressed concern that "where the need to meet federal diligent development requirements prevents coal producers from assembling the best possible packages of federal and nonfederal rights, the mining units which result may be costlier in acquisition, and less efficient in production, than those which could have been consolidated given a lengthier period of time."[26]

Another problem arises when federal coal must be formally committed to a particular project before work on the project can begin. The remaining time allowed under diligent development requirements may not be sufficient to complete very large and complex projects such as a synthetics plant. Coal industry representatives made the following complaint:

> A coal lease is only an initial step in a tremendously involved process of devising, financing and constructing a complicated production, marketing, delivery and utilization network for hauling the coal from a specific area to a

> specific plant. This is the way the system works. It has to work in that way. It can't be turned around. There is no logic in trying to draft regulations or enact federal statutes which require production from a lease before there are sufficient contiguous reserves, before the rest of the total system has been created and is ready to receive it, let alone before mine planning and environmental studies have been completed.[27]

The Rejection of the Free Market

Considering all the pros and cons of diligent development requirements, the dominant attitude in the Interior Department among those shaping the new federal coal-leasing program in 1974 was substantial skepticism. The most serious concern was that a diligent development requirement might frustrate, or at a minimum, significantly complicate the effort to construct a new federal leasing program based on market principles. The Assistant Secretary for Energy and Minerals, economist Jack Carlson, reflected the Department attitude:

> The introduction of requirements which cause coal to be produced at a rate faster than would occur in the absence of the requirement will usually introduce some degree of inefficiency into the production and consumption of coal. Since inefficiency means that the coal will not be produced and consumed at the lowest possible cost, anything which artificially accelerates the rate of coal production will cause the nation to waste its resources and eventually attain a lower standard of living than would otherwise be possible. The more diligence requirements tend to increase the rate of coal production beyond the desired rate of production, the greater will be the inefficiencies and the adverse impact on our standard of living in the long run.[28]

But the economics perspective, or for that matter any analytical considerations, could not really decide the diligent development issue. Whatever the analytical conclusion might have been, there was really no policy choice, at least in the assessment of top Interior officials. Strong public antagonism made it impossible for the Department to appear to countenance anything smacking of lease speculation; the Department could not overcome antipathies arising from a long history of conservationist and other opposition. The requirement for diligent development would prevent speculative holding of federal coal leases. Therefore, the Interior Department would have to have a tight diligent development requirement; it was as simple as that. There would not be a simple free-market solution to federal coal production.

5. A Planned Market Instead of a Free Market

There are many possible reasons why society might want to redirect the market. Perhaps the two most important are the cyclical instability of free markets and the distribution of income that results from them. In recent years the failure of traditional markets to account for environmental externalities might be added to these two.

Critics of the market often point to these failings as a reason for doing away with the market altogether. Instead, they propose government allocation of resources—the "direct command" strategy. On the whole, this view has dominated the implementation of environmental regulation in the United States in the past fifteen years. The Environmental Protection Agency, for instance, typically prescribes the specific control methods by which a mandated level of pollution emissions is to be achieved.

This is not the only answer, however. An alternative concept is that the market mechanism is still the best allocative tool available, even for achieving social objectives that must themselves be set outside the market. Indeed, in this way the best of both worlds is obtained. The great efficiency of the market mechanism in producing and distributing goods and services is preserved. Yet, society need not accept the verdict of the market with respect to ultimate goals; instead, market efficiency is utilized to achieve the goals which society itself selects.

Planned-Market Liberalism

The Brookings Institution has been a prominent designer and proponent of the modern welfare state in the United States. Yet, in recent years, it has emphasized the importance of using private market mechanisms for resource allocation. In *The Public Use of Private Interest*, published in 1977, a leading Brookings theorist, Charles Schultze, argued that by giving too many direct commands government was subverting many of its aims, causing wide public unhappiness: "The public has become disenchanted with the ability of government, especially the federal government, to function effectively. . . . The rash of new regulatory mechanisms established in recent years—for pollution control, energy conservation, industrial health and safety, consumer-product quality and safety, and the like—have generated a backlash of resentment against excessive red tape and bureaucratic control."[1]

However, Schultze did not see the solution in abandoning traditional welfare-state objectives. Rather, a better means had to be chosen to achieve these objectives. As Schultze states:

> There is a growing need for collective influence over individual and business behavior that was once the domain of purely private decisions. But as a society we are going about the job in a systematically bad way that will not be mended simply by electing and appointing more competent public officials or doing better analysis of public programs. . . . We usually tend to see only one way of intervening—namely, removing a set of decisions from the decentralized and incentive-oriented private market and transferring them to the command-and-control techniques of government bureaucracy. With some exceptions, modifying the incentives of the private market is not considered a relevant alternative. For a society that traditionally has boasted about the economic and social advantages of Adam Smith's invisible hand, ours has been strangely loath to employ the same techniques for collective intervention. Instead of creating incentives so that public goals become private interests, private interests are left unchanged and obedience to the public goals is commanded.[2]

The Schultze viewpoint is much removed from the old classical liberal argument that society is best off to leave the free market alone, that government intervention in itself is inherently undesirable. Schultze is, in effect, advocating strong government planning. But he wants a particular type of planning. Rather than direct government command, with its requirements for bureaucracy, the government should instead plan various levers and market incentives. If properly planned, such market manipulation will then induce the private sector to do of its own accord what the government actually desires. In this way government plans and policies can be implemented much more successfully through the time-honored tradition of voluntary market exchange; the major frictions and quarrels virtually inevitable in direct administrative commands by government to the citizenry can be minimized.

Such arguments for what I label in this book a "planned market" are a continuation of a tradition in which John Maynard Keynes is the dominant figure. Keynes sought to bring national product, employment, inflation, and other elements of aggregate economic behavior under government control. But he also sought to avoid what he considered the fatal medicine of direct interventions prescribed by socialism. By achieving government macroeconomic objectives by means of a few limited levers such as federal expenditures, taxes, and the budget deficit, Keynes sought to maintain voluntary exchange in the market as the basic mechanism for allocation of social resources. In 1947 one of his leading American disciples characterized this approach.

> The Keynesian economic system is essentially a machine which grinds out results according to where the several dials controlling the system are set. The functional relations are the building-blocks of the machine, and the dials are the parameters (levels and shapes) of these functions. . . .
>
> From Keynes' point of view the economic system . . . solved appropriately

> the problem of resource allocation; it failed only in its solution of the unemployment problem. The line of least resistance seemed, evidently, to be to improve the conditions of employment while still maintaining the capitalistic market mechanism for allocation of economic resources. Thus the Keynesian approach is clearly to modify capitalism so that full employment can be maintained. Any features of the capitalist system which do not interfere with the achievement of full employment may be preserved.[3]

In meeting other government responsibilities, such as environmental protection, society also must make plans and set certain targets that cannot be achieved by the market alone. But following Keynes, the government may be able to do this best by manipulating levers, such as pollution taxes, to achieve its plans, leaving underlying private market mechanisms undisturbed.[4] The creation of brand-new markets is another planned-market option—for example, a new market in pollution rights. Patents and other forms of property rights constitute additional tools available to market planners, which can be wielded to create the desired incentives.[5]

Although John Kenneth Galbraith is a major exception, the great majority of contemporary economists advocate in essence the creation of planned markets—even Milton Friedman. For instance, Friedman does not deny the necessity of active government intervention to achieve a more equitable distribution of income. However, he strongly advocates that it be done through the "negative income tax," rather than direct government redistribution of goods and services. Under the negative income-tax proposal (or the similar "family assistance plan"), the government creates a carefully planned labor market designed to supplement normal earnings while still maintaining work incentives and providing a floor living standard. Friedman's voucher proposal similarly aims at achieving a planned market in elementary and secondary education, replacing the current system of direct public school provision.[6]

The ideology of planned-market liberalism has a number of elements in common with old-fashioned progressivism. It is similar to progressivism in assuming that well-defined public objectives—in essence, a "public interest"—can actually be identified in advance and that politics and subsequent administration can therefore be separated. As in progressivism, democratic politics is to be limited to providing the broad social objectives; then, the planned market becomes the specific implementing mechanism free from political interference—the new analog to the old scientific administrators. Moreover, much as the science of public administration was to guide government actions to implement progressive social goals, economic science is to guide the government in planning markets to achieve welfare-state objectives.

The issue can legitimately be raised of the sense in which the ideology of the planned market constitutes a form of liberalism. Planned-market principles are, in part, an outgrowth of the old classical liberal tradition of free markets. However, they also take much from a contrary tradition which emphasizes the capabilities of government planning guided by scientific expertise—even if planning in a special market form in this case. As a practical matter, since the planned-market ideology is advocated by many "liberal" proponents of the current welfare state, and envisions an active planning and goal-setting role for government, it is considered below as a form of liberalism.

A federal coal program built on the prescriptions of planned-market liberalism would rely on specially designed market mechanisms to guide western coal development. The program for federal coal leasing has available a number of market levers with which to influence coal companies, including lease royalties, bonus bid payments, competitive bidding systems, minimum acceptable bids, environmental compensation and damage payments, and ordinary taxes. The objective would be to design the private incentives of the coal market so as to achieve the proper amount, location, and other aspects of federal coal production automatically. Environmental protection, for example, might be provided by requiring coal companies to make special payments, similar to pollution taxes, for specified environmental damages. Planning for federal coal would thus consist of planning the proper market forms, rather than planning the direct production of federal coal.

A Planned-Market Concept for EMARS II

The new Interior economists designing the program for leasing federal coal in 1974 were strong believers in the planned market, even if they would not have labeled it as such. One of them was Darius Gaskins, at that time Director of the Office of Mineral Policy Development. He was later to play a major role in the Carter administration deregulation effort as chief economist at the Civil Aeronautics Board and, following that, as Chairman of the Interstate Commerce Commission. The key EMARS II designer, however, was the Director of the Interior Office of Policy Analysis, William Moffat, who subscribed to the planned-market philosophy that "since we can't *order* coal companies to do what we want, and we don't intend to mine coal ourselves, our only option is to set up a process with regulations and incentives that will lead industry, acting in its own interest, to do what we think the Nation needs."[7]

The shift from a central planning to a planned-market approach to federal coal leasing was symbolized by the change in the name of the coal program from the Energy Minerals Allocation Recommendation System (EMARS I) to the Energy Minerals Activity Recommendation System (EMARS II). The Department did not want a whole new name, because it hoped to get away with writing a programmatic EIS in final form only, claiming the earlier draft EIS met the requirement for such a draft. The central role of market forces in EMARS II was later emphasized by Interior Secretary Thomas Kleppe when he announced the Department's new coal program: "The nub of the process . . . is the use of the marketplace itself, for the market will ultimately determine what coal, if any, will be leased and developed."[8] By mid-1974 and over the next twelve months Interior Department economists were actively engaged in the planning of new market mechanisms for federal coal management. One of the main such efforts involved the creation of private incentives to curtail speculative holding of federal leases.

Planning against Speculation

Once the Interior Department had concluded that a strict limit to speculative lease holding was a political necessity, if having little economic justification, it set about deciding just how soon development would be required and other details. The specific context was a policy discussion concerning the upcoming implementation of the longstanding, but never previously enforced, requirement for diligent development of federal coal leases.

The initial proposal, contained in proposed regulations published in December 1974, did not reflect the new planned-market philosophy. Instead, it prescribed a series of specific milestones which had to be achieved satisfactorily by a lease holder. Each lessee would have to show the government that he was actually "preparing to extract coal" by presenting evidence of mine plan preparation, environmental and geological studies, drilling, and other premining activities.[9] There were no definite time limits set, however; the ultimate determination of whether adequate diligence in preparing to mine was being shown was left to the local mining supervisor of the Geological Survey.

The public comments on the proposed regulations complained of their vagueness and the wide administrative discretion. On reconsideration, the Interior Department decided to adopt a new two-pronged system. On the one hand, a rigid diligence standard in which production must occur within ten years would be imposed. But on the other, market incentives for earlier production would be created through required payment of advance royalties, increasing each year from the sixth year onward. If production occurred later, the advance royalties would be credited against the royalties on actual production due at that time. If no production ever occurred, the advance royalty would be forfeited by the lessee. Hence, lessees with no intention to produce would have a strong financial incentive to abandon the lease earlier than the ten-year limit for getting into production.

The new system thus included elements of both a planned-market and a direct command approach. The Interior Department saw a key advantage in financial incentives in that they minimized the need for an intrusive bureaucracy:

> They usually impose no inspection or enforcement burden that would not otherwise exist—royalties and rents would be collected regardless. They leave to the operator the specific decisions on how best to adjust to the incentive, so they avoid heavy regulatory surveillance. And they can be designed to pro-

> duce smoothly graduated incentives, without the need for an either-or, yes-or-no decision by the government on whether the lease should be revoked; such yes-or-no judgments are difficult, because they make it impossible to allow for the fact that the degree of non-conformity may be either great or small.[10]

However, a fixed upper time limit was still considered necessary for political reasons. Moreover, economic incentives could not always be fine-tuned to achieve every desired result:

> Financial incentives, however, are difficult to design so that they accurately represent the government's motives, and so that they have no adverse side-effects. Advance royalties, for example, even if they are high, still permit speculative delay in production, because the profit from delay may be greater than the cost in interest on advance royalties. Further, such royalties may bias operators toward extracting Federal coal to the neglect of non-Federal coal. And advance royalties do little if anything directly to help increase the area of coal lands over which new environmental regulations could be applied, except insofar as they lead to relinquishment of existing leases.[11]

Publication of new proposed regulations in December 1975 stirred further comment; in particular, many objections were received as to the amount of time (ten years) allowed to get into development. The Peabody Coal Company, for example, made the following complaint:

> In many situations it is simply not possible to commence mining from an LMU within the period of 10 years from the date of the lease. A good example of this are Peabody's leases in Campbell County, Wyoming which are committed to a gasification project with Panhandle Eastern Pipe Line Company ("Panhandle"). The principal Federal lease involved became effective December 1, 1966. Since 1966 Peabody has been engaged in diligent efforts to develop this area. In 1972 an agreement was signed with Panhandle which should eventually result in a coal mine being opened. Since that date much development work has been accomplished. It appears, however, that a coal mine will not actually be producing coal until sometime in the 1980s. The point of this is to show that it is unreasonable to require that coal be produced from a Federal lease within 10 years. It may well be 20 years before Peabody produces coal from these leases in Campbell County, Wyoming despite diligent and continuing efforts by Peabody.[12]

Partly because of a large volume of similar objections, the Interior Department added provisions in its final regulations for the possibility of a single five-year lease extension beyond ten years.

Passage of the Federal Coal Leasing Amendments Act of 1976 necessitated further revision of the diligent development requirement. For any leases issued after 1976, extensions beyond ten years were no longer allowed. Moreover, the

Act included provisions limiting the ability to employ advance royalties, causing the Department to give up its effort to stimulate diligent development through such financial incentives. This result was apparently due mainly to an oversight in the drafting of the legislation—an example of the hazards of any kind of planning in the American system. The attempt to employ market incentives to control speculative lease holding thus was abandoned in the end. Such control would be achieved through direct government specification of an allowable time for development and through other commands.

Planning How Much to Lease

The requirement to develop a federal coal lease within ten years of acquisition meant that the incentives facing a federal lease holder with respect to the timing of coal production would differ considerably from those facing the owner of a private coal deposit. The private owner might well decide, given current coal demands and prices and the relative desirability of his coal deposit, to defer production into the future—say for twenty years. Thus, ironic in light of conservationist demands for a diligence requirement, this requirement might cause a federal lease-holder to begin production too soon, rather than conserve coal for the future. In the process inferior federal coal deposits might well be pushed into production ahead of higher quality private deposits.

There would be no problem if the federal government knew exactly how much federal coal, and which specific deposits, should enter into production under a socially efficient timing and spatial arrangement of western coal production. If it did, the government would simply lease this amount and those deposits. But it had no such detailed knowledge. It also could not simply lease federal coal greatly in excess of the amounts likely to be produced; such a policy would cause a distortion of production patterns in the presence of the diligent development requirement.

Moreover, a number of other practical considerations made it difficult to lease a lot more federal coal than seemed likely to be developed in the near term. Citizens of western states tended to equate the amount leased with the amount that would actually end up being produced. Leasing very large amounts of federal coal thus might well stir western fears of being overwhelmed with coal mining, possibly endangering the whole federal leasing program. At a less emotional level, western states also would be able to plan more successfully for coal development if they knew well in advance which specific federal coal deposits would be produced. Leasing large amounts of federal coal would entail a general loss of control over the specific configuration of federal coal development and the resulting environmental and other cumulative consequences. Leasing of federal coal well above production needs would also ensure that the Interior Department later would have to undertake widespread cancellation of leases that could not be diligently developed for lack of a market. This was likely to be a messy, time-consuming, and administratively burdensome task. Finally, leasing of large amounts of federal coal would encourage a new form of speculative holding of federal leases, with each purchaser gambling that his lease would be one of the select few actually developed in the time allowed. Such a result would be contrary to the original aim of the diligent development requirement to curb speculation.

Considering all these problems, but most of all the likely negative political reaction in the West to very large-scale leasing, the Interior Department resolved to try to lease an amount of federal coal reasonably close to the amount that would actually need to be produced. As the Department stated, "Leasing will be permitted only in amounts that are appropriate in relation to plans to produce the coal in the near future. The diligence requirements . . . are intended to insure this."[13] Interior Secretary Thomas Kleppe later confirmed a "guiding principle that only those deposits needed would be leased."[14] Following the basic concept of EMARS II, this objective would now be achieved by proper planning of market incentives: "The main policy change we've adopted is to recognize that under the old rules, firms were led to acquire huge reserves; under our new policy, the incentives cut the other way."[15]

Under the plan for the federal coal market the determination of the need for more federal leasing would be made on the basic concept that "the best way to tell whether a deposit of coal is 'needed' is to see whether the price at which it could be sold exceeds the costs (including environmental costs) of producing it."[16] The Interior Department asserted for this approach that "no other standard exists which so well applies a criterion of the public interest to leasing decisions. Leasing these deposits where value exceeds cost . . . will result in the proper overall rate of leasing, neither too fast nor too slow, and in fact is a much more certain way of arriving at the proper leasing rate than by central-planners' forecasts, which are notoriously subject to error."[17] Actually imposing a somewhat modified and stricter standard in practice (and economically necessary on various technical grounds), the operative rule would require that "Federal coal will be leased if an offer is made that is equal to or exceeds the estimated value of the resource deposit."[18]

To the predicted environmentalist charge that the new leasing program simply did the bidding of private coal companies, the response would be that:

> We project Federal production of perhaps 100-150 million tons per year by 1985, a large part of it from Wyoming, where the interest in production seems highest. But this is only a projection, subject to considerable error. Our basic answer to the question is: *no matter how much Federal coal is needed in future years, our procedures are designed to make that much, and no more available.* We're adopting procedures that do *better* than just make a central-planner's projection which is subject to error; . . . Our procedures will "meet the need" whatever it turns out to be.[19]

In essence, an automatic valve would be installed to regulate the flow of federal leases into the western coal market. As demand for coal rose, and coal company bids increased, more of these bids would exceed the fair market value requirement and more federal coal leases would automatically be issued.

In comparison with central planning, a key advantage of such a market mechanism would be the savings in information costs achieved by making individual decisions only as they arose concerning the issuance of leases. The administrative costs of a large government data gathering and forecasting effort for comprehensive planning would be avoided. Government would make much more effective use of industry and other already available sources of information. As the Interior Department summed matters up:

> . . . We have set a policy by which we will decide . . . case by case, and area by area, in the light of all the information available at the time. We have not decided in advance on an acreage or tonnage to be leased each year, but we have established a procedure by which a sound judgment can continuously be made as to whether the leasing rate is too fast or too slow. . . . And we will neither ignore the expertise of industry on demand, cost, and profitability, nor passively go along with whatever industry requests; we have instead established a procedure for combining information provided by industry with necessary added information on the environment and on local, regional and National interests.[20]

Fatal Flaws in the Market Plan

The total amount leased under the new coal program would be determined by the total number of bids received in excess of Interior Department fair market value calculations. However, this new procedure now placed a great burden on the fair market value estimates; indeed, setting fair market value now became the critical calculation necessary to achieving the right—the "needed"—amount of leasing. Despite Interior assertions, a new and very difficult planning problem, determination of the proper level of fair market value, had actually been substituted for the previous planning problem, the determination of the proper regional coal allocation. Moreover, there were two grave flaws in the EMARS II plan for the federal coal market, one practical and one conceptual.

A government requirement that a coal company bid must equal or exceed the tract value acts to limit levels of leasing in the following way. Some coal tracts will have a higher value if they are developed well in the future (even in "present value" terms). Such tracts may have poorer quality coal or higher mining costs than other tracts presently available. Because of their relatively lower quality, the development of these tracts will not become economic until the higher quality tracts have been exhausted. If current development would in fact be premature, then the government estimate of tract value should reflect the maximum value achieved by the tract at its optimal point for development in the future.

In bidding for federal leases, however, coal companies cannot bid more than the current development value of the tract. This is because, if they actually win the tract, the government's diligent development standard requires rapid development. In short, if the value of a tract is maximized with future development, coal companies bidding in current lease sales will not be able to offer the full value of the tract. The government therefore will find itself rejecting company bids for tracts best developed in the future. In this way federal coal leasing could in theory at least be limited to the amounts "needed" for development right away.

However, the implementation of such an approach would require highly sophisticated calculations by both government and industry. Furthermore, far from making such calculations, the Interior Department under EMARS II failed even to recognize the necessity for making them. Instead, in estimating the value of a coal tract, the Department never proposed to examine future development prospects. The fair market value of a tract would be simply calculated to equal the value of the tract if it were developed right away (or at least fairly promptly). No thought would be given to whether the tract might have a higher value if it were developed say 15 or 20 years from now.

Hence, despite Interior Department assertions, under the EMARS II approach every tract offered would have been leased, so long as government and industry made the same assumptions and used the same methods for calculating development value. The EMARS II plan offered no sure mechanism for controlling the level of leasing, other than limiting the number of tracts offered for lease initially.

In actual practice, bids for some (or many) tracts might well be rejected. But the pattern would be random; the rejected tracts would be those for which the government—for whatever reason—had formed a higher estimate of current development value than industry had formed. If the Interior Department were systematically calculating higher current development values in some lease sales, almost all the coal company bids might be rejected as insufficient in these sales. In other lease sales the opposite result might occur.

The EMARS II plan for the federal coal market thus was unsound in its basic logic. However, its conceptual defects could have been remedied by using the long-run value instead of current development value in forming government estimates of fair market value. But even if this concept had been adopted, there would still have remained another grave flaw in the market plan. The government in fact lacked the capability to calculate development value—either current or long run—with any real precision.

There are major practical difficulties in calculating development value of a coal deposit. This value depends on projections of coal prices, not only at the time development begins, but for periods extending up to thirty years. Costs must also be projected over the same long periods. An interest or discount rate must be chosen to convert future income streams into a "present value." The value calculation will be very sensitive to the choice of discount rate; for a coal deposit mined ten years in the future, a 10 percent discount rate, for example, will yield a value that is 37 percent less than a 5 percent rate. Coal companies do not all use the same discount rates, nor is the government necessarily well informed about such internal corporate matters.

The appropriate long-run coal prices to assume creates an especially big problem. In many calculations prices can be taken as given, determined by outside supply and demand forces in the market. But federal coal is such a large share of both western and national coal production that the amount of federal coal produced can itself significantly affect long-run coal prices. Hence, to estimate future prices for federal coal with any accuracy, it is actually necessary to perform essentially the same comprehensive balancing of present and future coal supplies and demands, region by region, that would be required under a direct government allocation of federal coal production. It would necessitate a full computer modeling of the coal market over the entire period of concern. Such modeling would require comprehensive surveying of available coal deposits and the ordering of these deposits by region from the lowest to the highest cost deposits. It would require close examination of utility and other purchasers of coal to estimate their demand at various prices. Coal transportation costs from one region to another would have to be factored in. In short, switching the planning focus from the amount of coal produced to the long-run price of coal is not much help; it merely looks at a different side of the same quantity-price solution. The same central planning calculations have to be made and the same planning costs incurred.

The Interior Department designers of the EMARS II coal-leasing program had hoped to avoid the need for such complex demand-supply modeling and the high information and planning costs it would entail. In practice, more feasible, if less theoretically correct and precise, devices would almost certainly have been employed under EMARS II—probably by inflating current coal prices using an estimate of a proper inflation factor. However, price forecasts derived in this manner would not be much better than a stab in the dark.

With all the uncertainties and necessary simplifications, the estimation of the true value of a federal coal deposit would inevitably be a very rough approximation. Indeed, a few years later, after an Interior Department task force on fair market value spent considerable time studying problems of revenue and cost estimates, it despaired of ever achieving any exact numbers:

> Regardless of the degree of excellence of the government's discounted cash flow coal lease rent-evaluation model, the correctness of the government's estimates of lease rents will be, at best, uncertain. This is because the size of the rents is very sensitive to the f.o.b. price resulting from the lease sale—and this price is very uncertain. Also, in computing rent you are finding a small difference between two large numbers which magnifies the effect of uncer-

tainty. Unfortunately, satisfactory f.o.b. price-prediction models do not exist and are unlikely to be developable since the instability of the estimate is inherent in the problem structure. Because of this, the government's estimates of lease rents could conceivably be off by an order of magnitude with only slight changes in input.[21]

If the design for a planned market in EMARS II had ever been implemented, it is likely that the Interior Department would have soon abandoned—or probably never started with—its proposed method of determining leasing levels. Instead, an informal government allocation scheme would probably have been worked out, using some quick-and-dirty, behind-the-scenes calculations of actual leasing needs. At least for a starter, these calculations would have gotten the leasing program underway. The objective probably would have been to lease enough coal to leave a large margin for error, to the extent that the public would tolerate this level of leasing. Then, future leasing levels would have been influenced by the rate at which recent leases issued were being brought into production, the size of bids at actual lease sales, the volume of coal company requests for new leases, and other signals of genuine market demand for new federal leases. Only in this sense would it be realistic to consider that EMARS II would be a market driven system.

The real alternative to such a strategy would be a similarly crude pricing approach.[22] The government might set some minimum price for coal leases and then see how many leases were being awarded. If the number of leases somehow seemed too many, the price would be raised. If it somehow seemed too few, the price would be lowered. Once again, some sense of the "right" amount of leasing would be required. This would have to be based on some loose idea of future coal production levels in western regions—necessitating a form of allocational calculation, however rough.

However the Interior Department proceeded, central planning calculations of some kind would have been unavoidable, although perhaps of a crude sort and out of public view. The initial level of federal coal leasing would very likely have to be set by a direct government determination; furthermore, interest group politics would inevitably end up having a major say in this.

Once the diligence requirement had removed the timing of federal lease production from the market, other matters had to be divorced from the market as well. The best efforts of Interior economists to devise a practical way to determine the "needed" level of federal coal leasing through a planned-market mechanism had not been successful. And the reasons in this case had not been political or bureaucratic. It was simply that no plan for the federal coal market could be discovered that would satisfy all the social objectives which were simultaneously sought. The key obstacle was the inability to make use of speculative incentives. Because speculation is in essence the market way of achieving conservation, no amount of ingenuity was likely to come up with a successful market plan in which the private incentive to conserve—to hold off production for the future—was precluded.

The Interior economists who designed EMARS II had predicted that the conservationist proposal to adopt central planning would prove wholly impractical and lead only to the pretense of such planning. In the end, however, their efforts proved subject to the same criticism. The Interior Department had not really designed a planned market; it had only adopted the pretense of one.

6. A New Economic Conservationism to Protect the Environment

The development of a federal coal-leasing program was influenced in many ways by the environmental movement. As examined in Chapter 3, many of the strongest opponents of federal leasing were motivated by fears about the impacts of widespread coal mining in the western states, thus far largely untouched by its impact. In response, the EMARS II coal-leasing program was designed with the "first Department goal, to ensure environmental protection," a goal that was to be "stressed throughout the leasing system."[1]

A considerable economics literature was available to the designers of EMARS II describing possible market mechanisms for achieving environmental protection.[2] Economists have long recognized that the desirable features of the market require that market prices capture all the beneficial and detrimental impacts of productive and consumptive activities.[3] As long as this is the case, maximization of profits will lead to maximization of net social benefits. Problems arise, however, when some benefits or costs are not captured in the prices of inputs or outputs—i.e., when "externalities" exist. A large economics literature analyzes the consequences of such externalities, and calls for the design of new mechanisms to bring external impacts back within the market system. The classic solution is for the government to impose a tax equal in magnitude to the external effects. More recently, the emphasis has shifted to redesign of property-right institutions to facilitate negotiations among parties previously outside ordinary market channels.[4] The simplest solution is to define new property rights that formally "internalize" the external impacts—such as creating a trading market for limited rights to pollute the air.

Until the late 1960s, such discussions were largely of academic interest. There were a few obvious and regularly cited examples of externalities, such as stockyards or a papermill. But economists commonly regarded the presence of significant external impacts to be an exceptional circumstance. However, the environmental movement changed all that. Environmentalists convincingly contended that pollution and other impacts outside the market were a pervasive element in modern life, to the extent that society's health and well-being could be threatened if pollution were not better controlled. Economists thus faced the task of translating their theoretical constructs into real world instruments of pollution control. Their efforts generally did not meet with great success. Congress and the executive agencies responsible for environmental protection generally opted for direct government commands in preference to market mechanisms of environmental protection.

Agency administrators responsible for environmental protection often started off with no great liking for market solutions; it was thus not so surprising that they did not turn to them. But this was not the case in the design of the program for leasing federal coal; in fact, it was just the opposite. Nevertheless, the results were about the same. The planned market was again rejected for direct command alternatives.

An early test of market planning for the environment involved a matter little known to the public, but in which a few coal companies had a major stake—the definition of "commercial quantities."

The Commercial Quantities Debate

As discussed previously, the Mineral Leasing Act of 1920 provides for noncompetitive as well as competitive coal leasing. It will be recalled that under noncompetitive leasing holders of prospecting permits can file an application for a noncompetitive (preference right) lease; the lease must be issued if a sufficiently valuable coal deposit (one having "commercial quantities") can be shown to have been found. About 180 such applications were still on file with the Interior Department in the mid-1970s, never having been processed because of the 1971 suspension of coal-leasing activity. The basic issue in the commercial quantities debate was whether coal companies would be able to convert these 180 applications into valid federal coal leases—which they could obtain without any competitive bidding or bonus payments. The environment was a concern in the matter because many prospective preference right leases were believed to be located at sites where unnecessarily great environmental damage might result from coal mining. Roy Hughes, Assistant Secretary for Program Development and Budget, argued that "it is inevitable, for example, that among the over 180 applications for noncompetitive coal leases, some are in areas where leasing should not take place because of environmental hazards. If we hope to avoid the potential environmental disasters that await us in these 180 applications, we must use the main opportunity we have left, and that is the valuable deposit or commercial quantities test."[5]

The specific question at issue was the legal definition of "commercial quantities" for the purpose of deciding whether to issue or deny preference-right leases. Assistant Secretary Hughes argued that not only should the coal have to be commercially minable ("marketable"), but it should also have to pass a broader test of social benefit, taking account of external impacts as well. It must be shown on "overall balance" to be in society's best interest to issue the preference right lease. "A deposit is valuable if the 'overall balance' of benefits and costs of producing it is favorable. The overall balance standard goes a step beyond profitability to the lessee, and includes identifiable benefits or costs to others as well."[6]

The proposed overall balance test was actually a bold challenge to traditional mineral law concepts. Going back to the mining laws of the nineteenth century, the decision to issue a patent or preference right lease had always been determined by the private economics of any minerals discovered. Now, the suggestion was made that the Interior Department should decide the question not only on the basis of mineral development value, but also any wider environmental consequences to society.

A planned market offered one means of achieving an overall balance of mineral and environmental values. Much like the idea of a pollution tax, the government would require payment for any damage done to the environment by mining federal coal. The required payments would be tailored to the specific effects of mining at each prospective site, representing an estimate—admittedly very rough in some cases—of the monetary value of damages done. In a simple case, a loss of wildlife habitat might be assessed at a certain dollar value of such habitat per acre—perhaps the cost of reestablishing it elsewhere. The surface losses of livestock forage or forest timber due to mining could be valued at their commercial market value without much difficulty.

If full damage payments were assessed, the government could then simply leave matters to the financial incentives it had created. If the applicant for a preference-right lease chose to commence mining in full knowledge of the damage payments he would be required to make, this decision would show that the overall balance of mining and environmental values compared favorably with other potential mining sites. Hence, the Interior Department could simply issue a preference-right lease automatically once production commenced (or some irreversible production commitment was made). If production did not occur (within some reasonable time), the application would automatically be rejected.

The Interior Office of Policy Analysis considered that such a market solution would be preferable: "The ideal case would be that of total self-administration, in which the government simply set the terms and conditions, and the applicant decided for himself whether to accept the lease." To be sure, this would necessitate actually requiring payment for harmful environmental impacts: "It would be logical to accept the applicant's judgment only if the terms of the lease included payment for the value of any remaining environmental damages. If such payments could be administered, the profit motive of the applicant and the overall-balance motive of the government would coincide, and the applicant's judgment could be accepted." The Office also saw clearly that, if the market approached were not followed, the alternative would be a direct bureaucratic command approach in which the Interior Department would have to make the decision itself on the actual net social benefits. "In the absence of such an ideal system of damage payments, however, it appears that the government would have to form its own independent judgment of the overall balance of favorable and unfavorable factors."[7]

A few years later, the Congressional Office of Technology Assessment (OTA) faced much the same issue and came up with basically the same analysis. It found that there was a fundamental dilemma in public minerals management. The government could not assert the right to stop development of new mineral discoveries without taking away much of the industry incentive to explore for minerals. On the other hand, the failure to establish public control left the possibility of major environmental damage. As a possible answer, OTA proposed for consideration a system based on financial payments for environmental damages set in advance. The first step would be a zoning procedure under which "surface use restrictions tied to land classifications" would be established by the surface management agencies as part of their normal land-use planning process—in essence, the analogue by zone of the individual lease stipulations imposed in the EMARS II system. Then, where damages were too expensive to control, "a schedule of payments could be developed along with surface-use restrictions as part of the land-use planning process for an area, with some nonmineral resource values being absolutely protected through restrictions and others being conditionally protected through compensation requirements. The individual explorer or miner could decide on his own whether the potential mineral values were worth the cost of paying for damage to the conditionally protected nonmineral resource values, and he could structure his project to minimize such required compensation by minimizing the damage."[8]

If the practical problems of setting environmental damage payments could somehow be overcome, OTA thought that this planned-market approach would represent the ideal solution, leading decentralized market forces to achieve a proper balance of economic and environmental values: "This option would replace the existing open-ended and broadly worded surface use regulations promulgated primarily at the national level with more specific and predictable conditions tied to land types and uses at the local level, substitute restrictions where appropriate, and ensure open access and secure tenure once such conditions and charges were firmly in place."[9]

But whatever the technical merits, the plan for environmental damage payments was fatally flawed politically. It required the Interior Department to put an explicit dollar value on the environment. In itself, such valuation would stir much opposition as a crude utilitarian affront to the morality of the new environmental or land ethic—as a "license" to pollute or otherwise do bad things. Nevertheless, there still might be some political willingness to prescribe environmental damage payments if they could be scientifically justified. But at the present time—and very likely for a long time to come—the payments would have to be set through a subjective judgment call of government officials as to the true monetary value of the environmental assets lost. Subjective judgments of this kind are made frequently in government or private life—e.g., How much is a government museum willing to pay for an artistic masterpiece? or, How much is an individual willing to pay for a book he would like to read? But government officials are very uncomfortable with formally specifying such values when no respectable scientific or technical defense can be given.

There is actually an asymmetry in the unwillingness of government officials to put a value on the environment. Officials have no special problem in answering the following question: Given a specific project with some financial value X, and an assessment of its environmental consequences, should development be allowed to proceed? An affirmative answer clearly says that avoiding the environmental damage is valued at less than this dollar amount. Nevertheless, political leaders will steadfastly decline to answer what would appear to be a very similar question: Given the environmental consequences, how much must the financial value of a future project be in order to go ahead with it? They could, for example, simply give the same answer X with no logical inconsistency; yet, top decision makers will regularly answer the former, but almost never the latter question.

One explanation often mentioned is the greater explicitness in assigning a precise value as opposed to an implicit valuation in approving a project. But probably more significant, political leaders prefer to minimize the number of difficult and controversial decisions they must make; they do not want to hand opponents an easy target or create new political foes unnecessarily. A decision on a specific project that is already proposed for some site may well be unavoidable. But to state how much a hypothetical project must be worth is to invite trouble for no necessary reason, and to bear a large potential political cost with little return. In short, while areawide advance specification of damage payments might have made good economic and market sense, it violated a basic rule of politics.

Such factors really precluded the use of damage payments from the start and they were never seriously proposed. If the goal of environmental protection was to be met in the processing of remaining preference right applications, some other administratively and politically more practical approach would have to be found.

There were in truth not many options. Ruling out the private sector meant that the government itself would have to do the balancing of development and environmental values. Even focused on single tracts, there was an unwillingness to quantify environmental values in monetary terms. Hence, the decision would have to come down to a discretionary call by the Secretary of the Interior concerning each preference right lease application: "This balancing decision, as in competitive bid evaluation, would not usually be a quantified dollar-against-dollar comparison, because the environmental considerations often would not be measurable in that way. Rather, it would be a judgmental decision, as at all other leasing stages where such balancing must take place."[10]

The large element of administrative discretion caused considerable unease within the Interior Department. The holder of an application might have spent a great deal on exploration, yet now have his lease application denied by the Secretary of the Interior—not because of any objective standard that he could have clearly foreseen, but because of an after-the-fact subjective call on the importance of damage to the environment. Moreover, the case-by-case balancing of development and environmental values at many individual sites might prove to be a major drain on the Department's administrative capacities. One Department official indeed feared that use of the overall-balance definition of commercial quantities "will, in my opinion, result in a large costly bureaucracy."[11] However, the Department was divided. Roy Hughes, Assistant Secretary for Program Development and Budget, was willing to bear the costs of the overall balance approach in order to achieve greater environmental protection. Most of the rest of the Department thought it impractical, administratively too burdensome, and/or unfair to applicants.

Assistant Secretary Hughes argued that the real issues were policy matters and not legal. However, the Solicitor proceeded to write an opinion that, legally, the Secretary would be on very thin ice with any other position than that "under the Mineral Leasing Act the Secretary has no discretion to refuse to issue a noncompetitive lease to a prospecting permittee who has made the necessary discovery."[12] Legal advisors often assume the key policy-making role in government because they give the final interpretation of vague and loosely worded statutes or regulations. The Solicitor's views in this case effectively decided the question. The overall balance test was, for practical purposes, dead after his opinion, although the final interment did not occur until later.[13]

A New Economic Conservationism

The definition of commercial quantities was the most controversial, but not the most significant, environmental concern in the leasing of federal coal. In the long run much more federal coal would be leased in future competitive sales than was contained in the backlog of applications for preference-right leases. Nevertheless, essentially the same obstacles stood in the way of a planned market to provide environmental protection in future leasing as well. Indeed, little consideration was actually given to such an approach. Instead, Interior economists moved in a surprising direction—towards their old foe, conservationism. But it was not the traditional Pinchot interpretation—rather, a new economically more sophisticated version.

Conservationism has been unfriendly to economic analysis for most of its eighty-year history—returning a mutual feeling. But more recently, the criticisms of economists—that conservationism is really a collection of pious platitudes, lacks any analytical rigor, and is really a pseudoreligious movement with its own "gospel"—have begun to hit home. In recent years, the Forest Service has brought in more economists and, on the surface at least, even begun to ask their advice on occasion. Partly in response, a number of professional economists have sought to show conservationists how economics can actually be introduced without doing serious violence to most of the basic tenets of conservationism. Indeed, these economists have asserted that a valid conservationism is possible only with such an economic foundation. The wide application of economic methods would finally provide the means actually to accomplish the scientific rationality that conservationists preached but seldom practiced.

The economic conservationist is distinguished from many of his fellow economists in his attitude towards the market. He does not feel as strongly that resource allocation is far better accomplished through the market mechanism. Indeed, if he is as much conservationist as economist—loyal to basic conservationist ideals, if not the specific means proposed in the past to implement them—he looks favorably on government. Conservationists have long shown an antipathy to private profit and instead have directed idealistic aspirations towards government and the achievement of "the public interest" by dedicated and selfless public service. Conservationism ultimately looks to government to solve the nation's problems, not primarily by manipulation of planned-market mechanisms, but by direct government assumption of responsibility. Thus, the natural response of the true economic conservationist to environmental externalities is not to look for planned-market incentives; rather, it is to redesign government decision making to account for the full social impacts.

Actually to put such conservationism into practice, it is necessary to define "the public interest." Unlike many others, the economic conservationist does not shrink from this imposing task; the definition shall, in effect, be "total value of national output of goods and services, including all environmental and any other nonmarket components." Operationally, to maximize this objective requires a set of dollar values for every good and service, market and nonmarket. For the marketable components of total national output, observed prices can simply be used—although there is a big problem here when it comes to long-run future prices. These are sometimes a key part of the calculation but they usually cannot be directly observed. Of the nonmarket components of gross social output, some will be similar to an ordinary marketable good or service—such as a day of hunting or a day at the beach. In such cases an equivalency formula to translate from market to nonmarket values can often be calculated without great difficulty. For many other nonmarket components, however, the job is not so easy. The true economic conservationist simply considers that it will often be necessary to exercise great ingenuity to overcome very formidable hurdles. What, for example, is the value of a clear view of a setting sun or of an endangered species that is saved?

The economist who has done the most in recent years to advance the operational techniques for valuing nonmarket outputs is John Krutilla of Resources for the Future, in Washington, D.C.[14] Krutilla is currently the leading economic conservationist in the United States. He has spent much time in proselytizing his economic interpretation to the bastion of old-style conservationism, the U.S. Forest Service—and with some success. Krutilla does not look to planned-market approaches, such as competitive bidding for use determinations on the public lands. Rather, he seeks direct government calculation of the specific inputs and outputs which maximize the net social value of the public lands.

In practice, Krutilla's proposals amount to the adoption by the Forest Service and other public land agencies of formal benefit-cost analysis as the primary basis for almost all their land-use decisions. Krutilla considers that "the principal goal of the Forest Service . . . is to manage the national forests and rangelands consistent with efficiency criteria for resource allocation in order to maximize benefits."[15] A critical first step will be to calculate the value of commercial and noncommercial forest outputs. "When we speak of multiple use where there are several different forestland products or services, the only way in which we can speak of maximizing the composite benefits is to use prices or implicit exchange values which offer a single, common, metric and allow weighted aggregation across all of the uses. This is the concept of economic yield or value, broadly defined."[16] It will require "very exacting work of a fairly specialized genre, in order to estimate the value, or benefit."[17] Krutilla considers that maximization of social value is the only defensible interpretation of conservationism. If it turns its back on economic criteria, the Forest Service will never be able to achieve a "coherent management rationale for multiple use, sustained yield."[18] In a speech to the forestry profession Krutilla delivered the message that the new economic conservationism actually "represents at once a most worthy, even if demanding, challenge and a great opportunity if the profession retains enough of its youthful spirit to respond creatively and vigorously as it did 'in the beginning.'" Foresters must take necessary if often unfamiliar steps to implement the precepts of economic conservationism; the profession must "send the planning and management staffs back to the professional schools to pick up the dimensions of competence currently lacking."[19]

Identifying Areas Unsuitable for Coal Mining

Blocked by the political and administrative hurdles discussed above, the designers of EMARS II never made much headway in looking for a suitable planned-market mechanism for environmental protection. But if environmental protection would have to be accomplished by direct central-planner command, at least the resulting planning decisions should be based on valid economic principles and logic. Hence, the designers of EMARS II had to turn to the economic conservationism advocated by Krutilla and other economists of this school. All environmental benefits and costs had to be factored in at all key coal-leasing decision points. The cost of coal mining must include not only the direct cost of extracting the coal from the ground but also the cost of all steps taken to protect the environment. In addition, the value of any surface uses lost during mining constituted a valid opportunity cost that had to be assessed and factored into federal coal decisions. The outputs of coal mining must include not only the coal itself but any environmental damages that could not be prevented. Instead of the private profit to the coal company, the net social value of coal mining, including the cost of any harm done to the environment, must be calculated as the basis for government decisions.

One of the first Interior Department attempts to put these precepts into practice was in a 1974 analysis by the Office of Minerals Policy Development (OMPD) of a proposed issuance of a phosphate lease in the Osceola National

Forest in Florida. The Interior Department had responsibility for this lease as the manager of federal minerals. The proper standard for lease approval was stated clearly: "Open pit mining of phosphates on a National Forest or elsewhere results in significant environmental impacts. The economic value of the Osceola Phosphates must be balanced against the economic value of the alternative uses of the Osceola National Forest as well as the environmental externalities which would result from open pit mining." The study then sought to determine the answer to the following question: "Will the benefits of mining be large enough to compensate for the losses of alternative uses of the mined area and the quantifiable environmental damages resulting from mining?"[20]

Examining expected future phosphate prices and outputs, the OMPD study estimated that the Osceola phosphates would generate economic rents having an expected value of $52.1 million. The estimate of the values of the land uses displaced at the site of the phosphate mining were: timber harvesting, $12.1 million; recreation, $1.1 million; and cattle grazing, $84,000. An extra expenditure of $889,000 would be required to replace local infrastructure. There would be 28,000 acres of wildlife habitat seriously disrupted, if not destroyed, as this was the home for several threatened and endangered species. Mitigation of the impacts on these species would cost $14 million. If the surrounding water table were depressed, and, as a result, contaminated by sea water, the cost of importing substitute fresh water would be $2 million. Finally, there was some chance—presumably thin—of radioactive wastes from phosphate mining entering the aquifer. The study concluded that there was no way either to estimate the probability of this occurrence, or of calculating the value of the damages that would result.[21] In sum, the total cost of the measurable foregone uses and environmental damages was $30.2 million, which was less than the economic rent of $52.1 million; hence, justifying the mining as far as the measured net social benefits.

The Osceola study was especially important because it turned out to be the model for an attempt to apply the principles of economic conservationism much more broadly to federal coal leasing. A 1975 memorandum from the Assistant Secretary for Program Development and Budget proposed a five-step system for "environmental balancing in mineral leasing," with particular reference to federal coal leasing.[22]

The first step was to exclude tracts for which "information on environmental values" was already sufficient "for purposes of ruling out areas where leasing should not be considered." This step, however, was only the initial rough cut. The Assistant Secretary stated, "We can use rather general environmental information, since we are trying only to rule out the very worst cases, and since we know that by the bid evaluation stage more detailed data will be available on the basis of which a tract can be withdrawn from lease if necessary."[23]

The second step would determine actions to be required in order to limit the environmental damage from coal mining—such mitigating actions to be specified in a set of stipulations put into each lease. However, it would be important to recognize that not all harmful impacts could or should be prevented. Mitigating measures would be justified only when the benefits of reduced environmental damage exceeded the costs incurred. As this principle was later explained: "Setting cost-effective terms and conditions has the important characteristic that not all damages will be stipulated against. The reason is that, typically, it will pay to prevent only part of them. The cost of prevention normally rises with its degree, and perfect protection against environmental damage is usually unreasonably costly. This means that a proper public-interest balance will be struck, in setting the terms and conditions of a lease, with some damages remaining."[24]

The third step would be the most important. An estimate of the private value of the tract to coal companies would be made, taking account of any private expenditures required to protect the environment. In genuinely competitive lease sales, the high company bid would be a good proxy for this value. A further estimate then would be necessary, of the value of environmental damages which were simply too expensive to prevent and thus would have to be borne. Situations might arise in which "after considering the cost of meeting all cost-effective stipulations, the lease would be profitable *for the operator*, but nevertheless the remaining damages would be great enough to make issuance of the lease *not* in the *public* interest."[25] Thus, the key test would be whether the development value of the tract more than made up for the unavoidable environmental damages. If yes, the lease could be issued; if no, it should be ruled out of consideration for leasing. Land-use maps would be prepared showing the areas in each category.

The fourth and last step, occurring after lease issuance, would be an environmental review of the mining plan when it was submitted to make sure that no major impacts had been previously overlooked and that lease specifications had been met. The full four-step process thus was designed to put into operation the basic standard that "a coal deposit should be leased and developed if the . . . *value* of the coal produced exceeds . . . *all costs* (including environmental and alternative land use costs) of producing the coal."[26]

The Actual Failure to Properly Account for Environmental Costs

In the Powder River Basin an acre of good coal land could easily contain more than 100,000 tons of coal. There are many complex technical issues in determining the true "social value" of this coal. But to simplify, the Interior Department will probably collect federal royalties of about $1.00 a ton. State and local taxes will also very likely exceed $1.00 a ton. Therefore, government alone will almost certainly receive total revenues of at least $200,000 per acre for many coal deposits in the Powder River Basin. Including all consumer and producer benefits (in technical economic terms, "surplus"), this is only a part of the total social value of the coal, and perhaps not the major part. Other western coal lands typically contain considerably less coal, but the basic point nevertheless is much the same: the value of a good western coal deposit is high, indeed almost certainly higher than any but the most exceptional of other uses of the land. Hence, in all likelihood very few coal tracts would have been ruled out by the criteria that were established under the EMARS II coal program.

But it would be incorrect to conclude from this that environmental concerns need not enter into decisions on where to lease federal coal. Much to the contrary, such concerns might indeed be the decisive factor. The real problem was in the incomplete way the EMARS II program had formulated the problem of environmental protection. Instead of absolute values, the emphasis should have been on comparing relative coal development and environmental values among different sites; instead of identifying broad unsuitable areas, the focus needed to be on selecting the few best areas or tracts for coal development from a very large universe of suitable coal-mining sites in the West.

There is enough demand for many federal minerals—such as oil and gas on the Outer Continental Shelf—that all the potential output of the federal lands could be sold in a short time without greatly affecting the market price of the mineral. But with 300 or more years of coal available in the United States, as much as one-third federally owned, obviously the great majority of federal coal tracts are not needed for mining right away. Hence, a coal tract should be developed only if it is among the very few tracts of highest quality that are actually needed for current development. The principle is the same as that under which oil shale and other relatively expensive oil sources will only be developed after other cheaper sources have already been exhausted.

The ranking of federal coal tracts should not be on the basis of private coal value alone, but of the net social benefit—the overall balance—reflecting environmental damages as well. The tract selection criteria of the EMARS II coal program, however, would not have achieved this objective. If literally applied, they would have recommended that almost all tracts be made available for lease sales. But once having made a large number of tracts available, the few tracts actually developed would then be decided by coal companies; private incentives would sort out the particular tracts with the highest private coal values of all. This priority ordering might differ substantially, however, from an ordering based on the social gains from developing each tract, since it would not take relative environmental impacts into account. In short, although EMARS II proposed to adopt the principles of economic conservationism with respect to environmental protection, it never proposed an effective means of real-world implementation for the actual key problem—selection of the specific tracts to lease. The problem on which EMARS II did focus satisfactorily was on areas wholly unsuitable for coal mining; but this only involved a small acreage.

Although the EMARS II coal program contained such deficiencies in its concept, it might have worked well under certain circumstances. As long as environmental costs showed much less variation from site to site than coal-mining costs, the main concern then would simply be to sort out the tracts with the highest coal value—something best left to the private market alone. EMARS II also offered a practical approach if all tracts happened to be similar both in coal value and environmental costs. If this *were* the case, the tracts actually developed would not make much difference and the EMARS II system would avoid substantial administrative burdens of government tract investigations. On the other hand, if coal value varied little from site to site, but environmental costs differed substantially, EMARS II would be the wrong approach. Relative environmental costs should be the basic determining factor in the final selection of tracts for development. Incorporating this factor would require direct government control to the very end of the selection process.

Still, even here, common sense no doubt would have prevailed. If some tracts had shown much greater environmental problems, they would very likely in practice have been excluded from consideration for leasing on whatever programmatic grounds were available. It might have been said that the absolute coal development value was less than the absolute environmental damages, even where the real consideration was simply that the relatively high environmental cost had to mean a relatively low priority for mining. In government, achieving the right results stands in higher regard than always having the logically correct reason—fortunately.

The EMARS II difficulties in putting planned-market mechanisms for environmental protection into operation illustrate a wider problem. If pollution taxes or other planned-market approaches cannot be made to work, government will be forced to make the final decisions itself. Then it will have to know both the development value and the environmental cost. When the final decision falls to government, it assumes a decision-making responsibility traditionally held by the private sector. For its part, the private sector loses its independence and becomes a provider of information, support staff, and analysis for government decisions which industry then must carry out. If the trend were to continue very far, it would constitute a major reordering of the traditional government and private sector relationship in the United States.

The major inadequacies in the carrying out of economic conservationism under EMARS II may well have resulted because its economist designers simply did not have the heart for the task. Really committed to the use of the market mechanism, they could not swallow the major central planning role for government to which the logic of economic conservationism led—even if it was based in much better economic thinking. Moreover, as a technical matter, economic conservationism would have been very difficult to put into practice. Few of the numerous environmental and other nonmarket values required had been calculated. Any serious effort to determine such values or to complete other requirements for the needed calculations would involve a small bureaucracy in itself. Such a direction hardly seemed appropriate to the basic EMARS II philosophy to put the main decision burdens on the private sector and to reduce the direct role of government.

III. Judicial Insistence on a Conservationist Federal Coal Policy

Introduction

The American system of government was designed to disperse power—a deliberate protection against the threat of government tyranny that was fresh in the minds of the founding fathers. Their commitment to a division of powers is seen in the constitutional status of the states as separate from the federal government; the creation of a bicameral federal legislature; and the independent existence at the federal level of strong judicial, executive, and legislative branches. The resulting bulwarks to loss of political liberty are not achieved without a price, however. Because of the diffusion of political power and responsibility, the policy-making process in American government often produces stalemate, or when it moves, it does so tortuously. One of the great obstacles to effective government planning is the number of independent players who must agree to any plan.

In the case of federal coal policy the legislative branch played a lesser role in the 1970s; the Mineral Leasing Act of 1920 remained the basic guiding statute, although some significant amendments were made in 1976. The western states most affected by federal coal policy also did not have a large impact, although this began to change towards the end of the 1970s as an actual resumption of coal leasing came closer to occurring. The key actors in developing a new coal program were the executive branch and the judicial branch. Federal coal policy making for much of the 1970s might be described as a negotiation—or confrontation—between these two branches of government. The prominence of the judicial role in federal coal policy making was not unusual during the 1970s; the prevalence of such a judicial role in fact spurred strong criticisms and counterattacks on the rise of an "imperial judiciary."

7. The Judiciary Debates Its Proper Role in Federal Coal Planning

Planning Utopianism

The market mechanism basically works through trial and error. Its path to social efficiency is the selection of the fittest. But this evolutionary process involving numerous bankruptcies and failed hopes is socially costly as well as individually painful. It is especially unsatisfactory in cases where the errors involve not only the failure of the individual entrepreneur, but major costs to the rest of society as well—taking an extreme example, an airline company for whom an "error" is a plane crash. It might seem far better, and indeed has long been the hope, that the efficient solution might simply be identified from the very start by the sustained application of human intelligence. In the modern age the methods of scientific reasoning have greatly expanded the scope and power of human thought, giving reason to believe that it can solve almost any problem. Those who have aspired to a less divisive, more humane, and also even much more efficient process than the survival of the few profitable, have usually placed their confidence in government planning. Planning has been the means of implementing faith in scientific rationality in social and economic affairs.

Expansion of government power has thus traditionally been closely linked to planning. More planning would necessitate bigger government, but equally important, the causation has worked the other way as well. There have been many groups in society unhappy with market results for whom the objective has been to achieve a greater government role. Yet, Americans have traditionally been very distrustful of government, fearing loss of political and economic liberties if government became too big and powerful. Government could be permitted to expand only if there were definite assurances that its new authority would be carefully controlled to prevent any arbitrary and capricious exercise. Fortunately, or so it seemed at one time, just this assurance could be provided by planning; a bigger government could be trusted because its exercise of power would be held strictly accountable to scientific rationality, as shown in formal plans. Such concepts were written directly into many of the laws which served to expand the role of government in the United States.

Zoning provides one of the best examples. A major progressive reform, zoning greatly expanded government control over the individual property owner. Zoning decisions could make some property owners rich while imposing large land value losses on others. Hence, the Standard State Zoning Enabling Act in 1924 specified that zoning regulations must be administered "in accordance with a comprehensive plan." When the Supreme Court approved zoning in

1926, despite widespread doubts that it was constitutional, the Court was much impressed that zoning was necessary to implement land-use plans to the benefit of the whole community. Regrettably, if necessarily, some individual property owners might on occasion have to suffer for that larger social good. In the following years state zoning laws generally directed that zoning should follow the guidance of a formal land-use plan. In the 1950s, a leading legal scholar of zoning went so far as to characterize the land-use plan as an "impermanent constitution."[1] Through court insistence, if necessary, the presence of the land-use plan assured that zoning would actually be administered to serve the public interest: "A plan should be more than advisory so that it may become part of the owner's arsenal of facts on which to exercise choices concerning land use and, more importantly, part of the material which a court may use to determine whether the administrator is acting in accordance with law."[2]

Although the role of government grew rapidly under the banner of rational planning, the reality of formal planning never came close to matching the wide hopes. Formal planning was another instance of progressivism showing lofty aspirations but much less followup on the details. An influential study of urban planning in Chicago during the 1950s concluded that "the idea of planning, or of rational decision-making, assumes a clear and consistent set of ends. The housing authority, we found, had nothing of the kind. The law expressed the objectives of housing policy in terms so general as to be virtually meaningless and the five unpaid commissioners who exercised supervision over the 'general policy' of the organization never asked themselves exactly what they were trying to accomplish. Had they done so they would doubtless have been perplexed, for the law said nothing about where, or in what manner, they were to discover which ends, or whose ends, the agency was to serve."[3] The National Commission on Urban Problems found in 1968 that "it could hardly be clearer that formal plans are not furnishing a unifying basis for most local regulations and other needed development guidance measures. Many communities regulate without meaningful plans. Some of those that do have plans pay little attention to them in making regulatory decisions. Still others find that plans provide little guidance in answering many regulatory questions."[4] Indeed, there has been a wide revolt against the very idea that plans can be "scientific." By the 1970s the leading theorists of planning were advocating a radical revision of traditional formal planning roles and ambitions: "In order to deal with their crucial problems, the cities will need not physical plans or comprehensive plans but plans that deal with specific problems and their causes, and that result in policies, or policy guidelines, action programs, and even individual decisions. . . . This requires planning analysts trained in new kinds of policy-oriented economics, political science, and sociology; and planners trained in policy formulation, program development, systems analysis and cost-benefit analysis."[5]

However, there was not a wide awareness among the general public (or even many policy makers) of the deep skepticism that was emerging with respect to comprehensive and other planning concepts derived from progressivism. Indeed, the emergence of the environmental movement at first produced an insistence on reviving just such planning—a demand finally to put into practice long-standing progressive and conservationist promises. The 1970 Clean Air Act Amendments and the 1972 Water Pollution Control Act Amendments contained major formal planning requirements. The Federal Land Policy and Management Act of 1976, the Federal Coal Leasing Amendments Act of 1976, the National Forest Management Act of 1976, and other public land legislation enacted in the 1970s also contained extensive requirements for formal planning.

As discussed in chapter 3, the basic strategy of environmentalists opposed to western coal development was to expose the wide gap between the ideal of scientific rationality which progressivism and conservationism offered, and the much different real world that was apparent in government decision making. Much of this effort went into exposing the moribund state of formal government planning—about which there was little doubt—and the failure to live up to earlier claims that comprehensive plans would closely guide government actions. Because formal planning was mandated in the law in many instances, those opposed to western coal development could legitimately complain that the government was actually violating obligations to which it was legally bound. However, there was little examination of whether such planning laws and expectations might have actually been impossible to satisfy in the first place. Indeed, if the real objective were to prevent most future development, the status quo was ideal; the government would always find it very difficult to live up to the planning preconditions it had imposed on itself for its actions.

Traditional conservationism never advanced much further in addressing practical problems of preparing plans than it did in explaining how economic tradeoffs should actually be made. Conservationist literature is weak on basic questions such as the desirable scope of plans, coordination among different plans, proper time horizons of plans, how specific or general the plans should be, how often plans should be reworked, and many other such real world concerns. Not only are these questions often not addressed, but in many cases they are never even acknowledged. A proper plan is said to be "comprehensive" and thereby the necessity for any limits to planning itself is denied. To any suggested gaps in proposed planning coverage, the rebuttal can always be made that the matter will be included, that the plan being prepared will genuinely be "comprehensive." A truly comprehensive plan, almost by definition, achieves all planning objectives. Comprehensive planning is really a utopian vision in the limited realm of planning, and one in which resource limitations do not exist.

The Problem of Planning Bounds

Like other forms of utopianism, claims to comprehensive planning can only be made beforehand; if anyone bothers to check afterwards, no plan actually completed can ever come very close to being truly comprehensive. Opponents of various government projects and policies, including western coal development, have been quick to see this, and to exploit the lack of any accepted principles of adequate planning scope. It will usually be easy to point out that important factors affecting the subject matter of a plan have been left outside the plan's scope, and thus it is not truly comprehensive. If a plan should then actually attempt to be more comprehensive by widening its scope, it will prove to be impossible to include as many details on specific actions or activities within the more encompassing planning bounds. Then opponents can criticize the plan's vagueness; there will be many lesser matters which the plan directly fails to address. This script has been followed in various policy controversies in recent years.

The National Environment Policy Act (NEPA) of 1969 was part of the rededication to formal planning stimulated by the early environmental movement. It is essentially a national planning act. Or, perhaps more to the point, because the language of the act itself is so vague, judicial interpretations have transformed NEPA into a national planning act. As one Supreme Court Justice, Thurgood Marshall, candidly acknowledged, "this vaguely worded statute seems designed to serve as no more than a catalyst for development of a 'common law' of NEPA."[6]

Under various judicial rulings, spurred also by regulations promulgated by the Council on Environmental Quality, NEPA seeks to impose on government decision makers many of the procedures of comprehensive planning. Inventories of data and other basic information must be collected; the relevant alternatives must all be identified; the social and economic consequences of each of these alternatives must be fully examined; and, from a comparison of all the alternatives, a decision is then made which must be formally explained and documented. All this is vintage comprehensive planning. Although NEPA supposedly deals with the environment, environmental and nonenvironmental factors are interwoven so inextricably that a satisfactory examination of environmental features requires that many other elements be included as well.

In transforming NEPA into a planning law the judiciary has had to contend with the question of the proper planning bounds. How comprehensive should plans be? What kinds of plans for what places should be legally required? Judges have had no greater success than others in resolving the problems of the proper scope for a plan. In fact, their efforts have been hindered by a lack of judicial familiarity with the history of planning theory and practice and by the special restrictions which the traditions of legal reasoning impose. Many judges may never even have perceived explicitly that they were engaged in the formulation of a new national planning law.

The federal coal program generated one of the major court cases concerning the scope for planning, which was eventually decided by the U.S. Supreme Court.[7] The specific issue was whether the federal government should be required to prepare a comprehensive development plan for the Northern Great

Plains, a region encompassing the Powder River Basin and other areas of widespread federal coal ownership in northeastern Wyoming, southeastern Montana, and western North Dakota. As mentioned earlier, this region has some of the richest coal deposits in the world—in large part federally owned—and may by itself provide as much as one-third of total U.S. coal production by the 1990s.

Sierra Club v. *Morton:* Judge J. Skelly Wright's Decision

The federal government possesses control over most key levers for the development of the Northern Great Plains. These include not only the issuance of federal coal leases in an area where most of the coal is federally owned, but also controls over much of the key infrastructure. Any significant new railroads, pipelines, highways, or other major facilities are likely to require a federal grant of right-of-way. The federal government also has exercised a dominant role in the construction and operation of the water supply system in the region. In short, as Judge J. Skelly Wright of the U.S. Court of Appeals for the District of Columbia described the situation, "Interior has power over development of projects in the Northern Great Plains that are very closely tied to any future development of the province's coal resources." As a result the future development of the Northern Great Plains was securely in the hands of the federal government: "Most, if not all, of the ongoing and pending actions in the Northern Great Plains are subject to federal approval and are thereby federal actions."[8]

Judge Wright considered the exercise of this control a grave federal responsibility: "The severity of the environmental effects of massive coal development of the Northern Great Plains is clear. . . . Briefly put, a region best known for its abundant wildlife and fish, and for its beautiful scenery, a region isolated from urban America, sparsely populated and virtually unindustrialized, will be converted into a major industrial complex." He found that "the spectre of significant, and unnecessary, harm to large tracts of valuable wilderness still remains." Given these circumstances, it was incumbent on the federal government to plan carefully for the future of the Northern Great Plains. In fact, the Interior Department had already agreed with this assessment: "An assurance of orderly and timely development would require an analysis of such items as regional coal demand and the relationship to existing leases. The Department is initiating a State, local and Federal program to develop a regional development plan or framework for the Montana, Wyoming, North Dakota and South Dakota areas associated with the Powder River and Fort Union coal formations. The objective is the wise development of the region accomplished in full realization of social, economic and ecological consequences of alternative possibilities."[9]

The Interior Department had already sponsored a series of major studies of the Northern Great Plains Resources Program, which was intended to coordinate federal activities in the region and set a policy framework for future management decisions. Interior Secretary Rogers Morton said, with regard to the study, that "the vast reserves of coal . . . provide an excellent opportunity for this Department to demonstrate how a responsible Federal agency can manage resource development with proper regard for environmental protection. It is important that we not lose this opportunity by engaging in single-purpose studies which are incapable of developing comprehensive information or by taking piecemeal actions which restrict our future options."[10] The Interior Department announced that for the duration of the study it would cease approving mining plans, granting coal related rights-of-way, authorizing new delivery and sale of water, and issuing new coal leases or special use permits. In short, beginning in 1972, the Interior Department embarked on what certainly looked to be the preparation of a plan for the future coal development of the Northern Great Plains.

Watching this undertaking, the Sierra Club concluded that any plan for the development of the entire Northern Great Plains must certainly be a major federal action. It therefore filed a suit to require preparation of an environmental impact statement (EIS) for the plan. Although the complaint was denied in the Federal District Court, on appeal to the D.C. Court of Appeals, Judge Wright joined with another colleague in a two-to-one decision which sustained much of the Sierra Club complaint and enjoined almost all further federal coal actions until the matter could be remanded to the lower court and certain remaining issues resolved there. Judge Wright determined that "it is our view that when the federal government, through exercise of its power to approve leases, mining plans, rights-of-way, and water-option contracts, attempts to 'control development' of a definite region, it is engaged in a regional program constituting major federal action within the meaning of NEPA." Moreover, it was not very important that the Interior Department now denied that a plan had ever been intended; the existence of a plan depended on the effect of its actions and the way they were carried out, not "whether it labels its attempts a 'plan,' a 'program,' or nothing at all." The courts, if necessary, should not shy away from "*imposing* a requirement of comprehensive planning on the government when it refuses to do so itself."[11]

Judge Wright carefully developed the logic of comprehensive planning. He noted that the courts had already determined that individually minor federal actions could still have a "cumulative effect" which "could constitute a major federal action" and thus require an EIS. In short, planning clearly could not stop at the bounds of each minor action; rather an overall plan encompassing the minor actions might well have to be prepared. In the case of the Northern Great Plains, most of the individual actions were already significant enough by themselves to qualify as major federal actions requiring an EIS. Judge Wright then concluded that the "cumulative impact" of "admittedly major federal actions" could also require an EIS. Even plans that already had a fairly wide scope thus might have to be brought together under a still wider plan encompassing them all. Courts thus might find that "agency violation of this substantive duty by a failure to improve its plans or coordinate its actions might justify a judicial directive to coordinate various major federal actions into one comprehensive major federal action, followed by a directive ordering issuance of a comprehensive impact statement for that newly-comprised action."[12]

However, Judge Wright was not so taken with the logic of comprehensive planning that he failed to notice that it could lead to wholly unreasonable conclusions, or as he understated the matter, that there were "practical difficulties in its broad application." Since interdependencies exist at some level among all elements of the social and economic system, what would limit the requirement for comprehensive planning? "An infinite number of geographic, environmental, or programmatic interrelationships might be found among the various individual projects under way throughout the country." But "surely . . . an infinite number of comprehensive plans, and comprehensive impact statements, are not required."[13]

Judge Wright recognized that no clear objective basis seemed to exist for deciding plan bounds. The agencies had to be left considerable leeway to make this call: "It is, of course, the agencies that are supposed to organize the various federal projects throughout the country, not litigants who come into courts seeking, possibly in opposition to one another, to force the agencies into action according to their lights." If they sought to prescribe plan bounds, the courts might become involved in a morass of administrative detail: "Use of NEPA to force a comprehensive plan on an unwilling agency as a means to force that agency to undertake a comprehensive impact statement might intrude unduly on agency discretion, while overly involving the courts in the day-to-day business of running the Government." In the particular instance at hand, "one might seek, for instance, a coal impact statement for the Northern Great Plains, another for the four-state region, and another for the State of Wyoming alone."[14]

The number of plans that should be prepared depends on the cost of planning. If planning were costless, very large numbers of plans could reasonably be undertaken; a plan could be prepared whenever any significant social or economic interrelationships seemed to exist. It is because plans are often expensive that difficult decisions are required as to how many plans to prepare and what the proper plan bounds should be. The cost of the plan includes the direct expense in manpower and material required for plan preparation. But another significant cost of planning is often the cost of awaiting plan completion. Hence, there is always a need to ask questions such as: Has the proposed action already been reasonably well justified? How much difference would it make to take more time to consider the matter? Have the purposes and goals been thought through methodically enough? Are there real major alternatives not yet adequately considered? In brief, have we done enough planning? There must be a carefully developed plan for the act of planning itself.

NEPA has greatly enhanced the power of the judiciary to make a final cut on the benefits and costs of further planning—i.e., to become the chief planner for planning. The problem is that the answer is generally a highly subjective judgment. Moreover, judges have no special qualifications in evaluating either the benefits of further planning or the costs of delay.

In his decision Judge Wright's only comment about the cost of delay was the cryptic remark that "federal approval of development of the Northern Great Plains seems fairly certain to occur, and to occur in the relatively near future so as to lessen our dependence on imported oil." As far as the benefits of delaying matters to allow further planning, he suggested that this depended on: "To what extent is meaningful information presently available on the effects of implementation of the program, and of alternatives and their effects? To what extent are irretrievable commitments being made and options precluded as refinement of the proposal progresses? How severe will be the environmental effects if the program is implemented?"[15] On these grounds, to which he devoted a couple of pages of discussion, he found the potential for significant benefit in preparing a plan for the whole Northern Great Plains.

The NEPA legal context, in which a decision on the need for further planning was made, was not conducive to a careful analysis of the question. An agency examining the benefits and costs of a water project or other facility would never think to rest its case on the few skimpy details and thinly supported assertions that Judge Wright employed in analyzing the benefits and costs of further planning in the Nothern Great Plains. Certainly, after NEPA no court would have accepted this standard of analysis for an environmental impact statement. Yet, the need is as great to plan for planning as for any other action—perhaps even more important.

In his memoirs Supreme Court Justice William Douglas commented approvingly on a candid remark of the former Chief Justice Hughes that "ninety percent of any decision is emotional. The rational part of us supplies the reasons for supporting our predilections."[16] Indeed, the most likely explanation for Judge Wright's injunction on any further coal-development activity came at the end of this decision. Sounding very much like a latter-day Gifford Pinchot, he acknowledged his strong personal sense of the urgent necessity to make "environmental planning a day-to-day occurrence," which "will make the needless abuse of our priceless national heritage a nightmare of the past."[17] It had long been a main article of conservationist faith that planning was the means whereby the public interest was asserted against the special interests, and rational social resource allocation was achieved. Judge Wright proved to be a true believer. His desire to reaffirm the conservationist gospel exceeded the reservations he showed about its practicality.

Sierra Club v. *Morton:* The Supreme Court Decision

The U.S. Supreme Court, however, proved not to have the same faith as Judge Wright. Its own nightmare was an endless series of plans and continual controversial appeals to the courts to decide where and when land use and other plans should be prepared. Writing for the Supreme Court, which unanimously overturned the Appeals Court decision, Justice Lewis Powell indicated his difficulties in specifying any clear basis for an appropriate planning scope.[18] He observed that the Interior Department was already committed to preparing plans of lesser but still regional scope—a "regional EIS" whenever "a series of proposed actions with interrelated impacts are involved . . . unless a previous EIS has sufficiently analyzed the impacts of the proposed action(s)." Justice Powell reasoned that it was not up to the courts to second guess the agency choice of regions, unless it appeared clearly arbitrary: "The determination of the region, if any, with respect to which a comprehensive statement is necessary requires the weighing of a number of relevant factors, including the extent of the interrelationship among proposed actions and practical considerations of feasibility. Resolving these issues requires a high level of technical expertise and is properly left to the informed discretion of the responsible Federal agencies."[19]

The Supreme Court did acknowledge the possibility that "comprehensive impact statements" might be needed. Such an EIS might be prepared, for example, where there existed "several proposals for coal-related actions that will have cumulative or synergistic environmental impact upon a region." But it emphasized that "determination of the extent and effect of these factors, and particularly identification of the geographic area within which they may occur, is a task assigned to the special competency of the appropriate agencies."[20]

The court also based its decision on other factors, in particular the fact that the Interior Department considered its undertaking for the Northern Great Plains a "study" and contended that it did not constitute a "plan." By this time the Department understood the legal importance in all of its statements to deny that any plan had existed or had been intended. Another influence on the Supreme Court was that the Northern Great Plains resource study, the alleged plan, had not yet been completed and thus the issue involved a matter only being "contemplated." The court, however, stated that "a court has no authority to depart from the statutory language and, by a balancing of court-devised factors, determine a point during the germination process of a potential proposal at which an impact statement *should be prepared.*"[21]

Sierra Club v. *Morton* was a landmark court case in the history of planning in the United States. If the Supreme Court had accepted the arguments of the D.C. Court of Appeals, the way would have been paved for the courts to order the preparation of plans of all kinds, whenever a court found significant interconnections to merit it. The Supreme Court, however, decided to beat a quick retreat from this potentially huge responsibility. The courts had already bitten off a great deal—perhaps more than they could really handle—simply in requiring acceptable plans (i.e., EISs) for individual federal projects or aggregations of projects clearly identified as bound together in a single program. If they had to examine every federal land-use or other formal plan for adequacy, or to redefine the proper scope of existing planning efforts, the work load might be overwhelming. Moreover, the courts had no special competency for this particular task.

The prospect of judicial involvement in economic and government planning was in itself no great problem. In fact, to the contrary, the judiciary has really been the closest approximation in the United States to a true national planning system.[22] The courts have had the independence from ordinary politics required in traditional planning theory. They may have lacked expert skills of a scientific nature, but much of their planning has not required them. Many court decisions concerning economic matters have really consisted of the design of a planned market. Thus, until the New Deal, the courts struck down various government efforts to set constraints on the market. Contract, patent, land-use, and other areas of law deal with the rights and obligations created among market participants and seek to structure them to achieve broad social purposes. Court decisions have defined the powers of regulatory agencies to intervene in the market. Indeed, the planned market offers wide scope for judicial economic planning for precisely the reasons that it appeals to economists. One need only design a few levers or controls properly; the market mechanism will then achieve the desired social objectives. The judiciary is well suited to focusing its attention on such narrow questions which still have far-reaching consequences. By contrast, the review of the adequacy of a comprehensive central plan would be wholly beyond the courts. One of the important consequences of a trend towards wide central planning would probably be a significant curtailment of the major judicial role in economic planning which is possible under planned-market liberalism.

8. The Interior Department Tries to Resume Leasing

The February 1973 announcement by Interior Secretary Rogers Morton of the Department's intention to develop a new program for federal coal leasing had stirred a flurry of activity. From 1973 onward leasing of federal coal would rank with federal leasing of oil and gas on the outer continental shelf and the future of federal lands in Alaska as one of the three most important and controversial issues before the Interior Department. A series of secretarial option papers had been prepared in late 1974 and early 1975 and decisions made on a number of key features of the new coal-leasing program. By mid-1975 the EMARS II program had pretty well taken shape and the question was how to announce it to the public.

A number of the most important program decisions have already been discussed above. The Interior Department would not try to forecast national energy requirements and then to allocate the overall forecast into regional coal requirements and then further into federal coal required within each region. Instead, the emphasis would be placed on signals received from industry showing the need for new coal supplies, to be indicated through industry nominations, the size of bids, and in other ways. Diligent development of federal leases for the first time would be strictly required, both for existing leases and for new leases. Advance royalties after the fifth year would smooth out the process of enforcing the diligent development requirement and create a financial incentive for early lease relinquishment. A requirement that, once started, leases be maintained in continuous operation would also be enforced, although the requirement could be relieved by payment of advance royalties.

The new program would also give a high priority to environmental protection. The BLM land-use planning system would examine environmental impacts of coal mining and assess the land uses that would have to be foregone if coal development occurred. Federal coal leasing would take place "within areas derived by the land use plan as suitable for coal leasing."[1] Within such suitable areas the tract selection process contained in the land-use planning system would then ensure that the coal value of any tract offered for lease exceeded its environmental cost plus forgone use value. Finally, all leases issued would contain stipulations and other requirements providing for environmental protection of the lease site during coal mining and reclamation after mining was concluded.

Royalty rates were sharply increased under EMARS II and were altered from a fixed payment per ton to a percentage of the coal value—8 percent for surface-mined coal and 5 percent for underground coal. All tracts were to be sold by competitive bidding, making permanent the existing moratorium on preference-right leasing (this decision was not finalized until somewhat later). Proposed coal lease sales would be analyzed in regional environmental impact statements; in effect, these regional EISs would constitute the formal lease sale plan. Regional EISs could also cover mining plans, rights-of-way, and other coal-related actions in the region. The boundaries of the regions would be chosen, taking into account factors such as coal fields, watersheds, common transportation routes, areas with close socioeconomic interconnections, political boundaries, and other types of interdependent features.

The Department decided to steer clear of the legal and administrative tangles in trying to assess an overall balance of coal values and environmental costs in processing the existing backlog of applications for preference-right leases. However, it tightened the standard for obtaining a preference-right lease, requiring that the applicant make a detailed and convincing demonstration of current commercial feasibility—the ability to make a profit from mining today.

Although the final decision on a few matters was not made until late 1975 or early 1976, most had been made by the summer of 1975 and had already been or could easily be incorporated into a formal statement for the Department's coal-leasing program. It will be recalled that the original leasing program, EMARS I, had been described in a draft programmatic EIS released in May 1974. Although this EIS had been severely criticized for its vagueness, given all the decisions just described, a much more detailed leasing program could now be presented. Additional material had also been added to the earlier EIS text concerning the importance of western and federal coal in meeting national energy needs and other improvements to the EIS made as well.

The Interior Department knew that while it probably faced a legal challenge no matter what it did, the challenge became a virtual certainty if it proceeded directly with a final programmatic EIS for coal leasing without first issuing a new draft. The Natural Resources Defense Council had sent a letter to Interior Secretary Morton in February 1975 saying:

> As set out more fully in our detailed comments on the draft EIS . . . NRDC believes that the draft was of such poor quality as to preclude meaningful comments. This view is consistent with the very critical comments you have received from various federal and state governmental agencies and responsible public interest groups. The Environmental Protection Agency, for example, gave the draft EIS its lowest rating ("inadequate"). The Institute of Ecology (T.I.E.) documented its shortcomings in considerable detail, as did agencies from the various states which could be affected by a renewed federal coal leasing program.
>
> The new EIS will undoubtedly be so substantially rewritten as to bear scant resemblance to the draft issued last year. We thus feel very strongly that a new *draft* EIS must be released and circulated for comment. Only in this way can the Department's credibility be preserved as it moves to formulate a coal leasing program, and the legal requirements of the National Environmental Policy Act of 1969 be met.[2]

A threat to sue from NRDC was no idle matter. NRDC had just won a suit against the entire BLM livestock-grazing program and the Department was being forced to agree to prepare 212 separate environmental impact statements stretching over thirteen years that might ultimately cost several hundred million dollars. NRDC was by a wide margin the most successful environmental litigant against the federal government in the 1970s, winning numerous major cases. The organization had been founded by Yale law school graduates and was supported by the Ford Foundation. Several NRDC staff attorneys had been law clerks for Supreme Court Justices. Its staff attorneys in an earlier era would have been bright young lawyers of the New Deal or corporate advisors in the best law firms. In short, the legal artillery NRDC could bring to bear was formidable.

Discussions within the Department concerning whether to issue a draft and then a final EIS for the EMARS II leasing program, or simply go directly to a final EIS, revolved around several key considerations. What was the urgency of a rapid resumption of federal coal leasing? What was the likelihood that the almost certain legal challenge to a final EIS would be successful? If the Department issued a new draft and then a final EIS, would the later final EIS have any greater prospects of withstanding legal challenge?

By the mid 1970s the writing of environmental impact statements had become a central concern throughout the Interior Department. Many Department projects and programs stopped or started according to the fate of their EISs. As much thought and effort now often went into the EIS strategy as the design of the proposed project. "Telephone-book" EISs became commonplace as Interior agencies sought to cover every conceivable subject that might cause an EIS to be rejected by a court. Such a rejection could easily hold everything up for several years. EISs began to consume large amounts of agency money, in addition to personnel.

EISs were in fact having a significant impact on the Interior Department in promoting greater attention to environmental protection, reinforcing various other pressures in this direction. But the reason was not the one given in the law. Few Interior EISs ever discovered any very surprising or unexpected impacts of projects on the environment. Rather, the EIS requirement gave environmental organizations and the judiciary a powerful new handle to influence Interior decisions. The judiciary often showed little restraint; for a couple of years in the late 1970s the Interior coal, rangeland, and timber management programs—a major part of the Department's overall responsibilities—were operating under court orders imposing significant management directions. Many Department meetings were preoccupied with EIS matters: How could the Department regain its former decision-making role? What concessions would be necessary to placate environmental opposition to prevent a suit or to gain a favorable court verdict on the "adequacy" of an EIS, allowing an Interior project to proceed? What were the tradeoffs between a weaker EIS right away versus a better EIS and resulting delays?

The Benefits and Costs of Delaying EMARS II

The importance of rapidly resuming federal coal leasing was the subject of considerable debate in the Interior Department. There was a strong desire to move ahead with federal leasing, especially since it had been stopped since 1971. Four years seemed a long time, particularly when a year and a half previously the first OPEC price hike had hit, and the nation now was anxiously searching for ways to reduce dependence on foreign oil. Western coal was one of the major untapped sources of energy; at a minimum, renewed federal leasing would have important symbolic significance as a demonstration of the nation's resolve to deal with its energy problems. Yet there were already huge supplies of previously leased federal coal available—far more than would be required for any conceivable level of western coal development in the near future. Moreover, the 180 or so backlogged applications for preference-right leases contained additional vast coal reserves, much of which were likely to end up under lease.

The importance of new federal coal leasing thus was clearly not a matter of any absolute shortage of available federal coal. Nevertheless, not much was known about the coal already under lease. Much of it had been leased a long time ago in a haphazard fashion. The odds were that some—maybe even most—of it was not the best federal coal available. Previously leased federal coal might not compare favorably with other federal coal in terms of low mining cost, minimal environmental damage, or high coal quality (sulfur content, heat content, etc.). It might be poorly located with respect to transportation or require joint development with unavailable adjacent private coal. These matters were not something that could be proven to everyone's satisfaction; rather, it simply stood to reason that many of these problems existed.

In summing up the importance of resuming new federal coal leasing, John Whitaker, Under Secretary of the Interior, pointed out that "despite large amounts of Federal coal already under lease in relation to production, because coal transportation costs are high and therefore coal demand is often localized, Federal deposits not under lease will in some cases be the lowest cost source of supply. This is true despite the fact that too much Federal coal may have been leased elsewhere. Failure to lease such lowest-cost deposits would raise total energy costs." This consideration was made more urgent because "Federal and non-Federal coal lands in many areas are intermixed in such a way that if the Federal deposits are not made available for lease, it may be difficult to put together mining operations of efficient size, and neither the Federal nor the non-Federal deposits will be developed."[3]

Whitaker recognized, however, that the case for renewed leasing was not an unarguable one, that he was simply making a "logical case for renewed leasing," not one based on any precisely demonstrated empirical necessity. Moreover, while the case was no doubt logically valid, it could not be denied that there was still a large amount of federal and nonfederal coal already available; Whitaker was not so sure that mining and environmental costs differed all that much from one coal site to another. In fact, he acknowledged that there probably was already a "wide availability of other coal deposits at costs near those of present mines." Given that "large amounts of non-Federal coal are available," he suggested that "postponing Federal leasing may create higher energy costs in some areas, but over all, since Federal coal is a small part of total production, the adverse effect would not be great for at least several years."[4]

While a delay in leasing might not be especially harmful, the question still had to be asked whether there was much to gain in using such a delay to issue a new draft programmatic EIS for EMARS II. Environmental organizations had strongly criticized the original draft programmatic EIS for EMARS I because the program description was so vague and brief that the public literally had almost nothing to review. The Interior Department basically agreed with this criticism and, as described, had spent the next year in high-level effort to flesh out a sensible program. While a few details still remained to be settled, the Department had made up its mind on most of the program by mid-1975. The Department thus had to ask: how much change in the federal coal-leasing program was likely to result from an improved programmatic EIS?

The Department took a narrow view of the function of environmental impact statements; they were literally to be statements of impacts. Although many observers argued that the impact statement should be the basic decision-making vehicle for putting together the entire federal coal-leasing program, the Department had rejected this interpretation of NEPA. Many of the new coal program details—rental and royalty rates, diligence and continuous operation requirements, bidding systems, fair market value estimates—had only minor environmental impacts, if that. Other aspects of the program with greater environmental consequence—tract selection and determination of areas unsuitable for federal leasing—were embedded in the BLM land-use planning system that was already five years old. The planning system also had few fixed standards that might be scrutinized and instead emphasized wide field discretion in land-use decisions; thus further studies in this area seemed unlikely to produce any major improvements.

There were some areas where a significant environmental impact would occur and Department policy might be altered. For example, not much attention had been given to the treatment of existing leases and preference-right lease applications in locations now judged to be unsuitable for coal development. But, on the whole, the likely improvement from further review of EMARS II did not seem very great. Barring a major change in program philosophy, which the environmentalist opposition admittedly was seeking, an extended EIS review probably would only be going through the motions.

In any event, the real question was not the structural details of the EMARS II leasing program, but whether the Interior Department should seek to encourage further western coal development by leasing more federal coal. As indicated above, the Department recognized that its case for renewed leasing might seem rather thin. On the other hand, the following question had to be considered: What would be required to demonstrate the need for leasing?

The General Accounting Office at the time was a leading critic of Interior Department efforts to move forward with federal leasing. It contended that the key issues of the leasing program should be fully resolved before lifting the moratorium on leasing. The GAO concluded that

> we believe that Interior should have as clear a conception as possible of the potential contribution of Federal lands toward meeting the national coal production goal. We believe it is not sound policy to begin a new leasing program without having reasonable goals of how much coal to lease and when to lease, based on the best possible estimates of how much coal to expect from development of the leases.[5]

Environmental organizations had similar views, as indicated in congressional testimony of a leading spokeswoman. First, federal leasing should not be resumed just because the Department had a suspicion that unleased federal coal might be superior to already leased federal coal. Rather, the Interior Department was under the obligation to prove its case. It must gather "the necessary information on which to make decisions" and must accept "the burden of proof . . . to demonstrate . . . that the lease is needed." By contrast, EMARS II was "long on justifications as to why still more leasing is necessary, and short on data to back up their claim." For example, "one justification they have put forward is that vast amounts of leased coal is 'unavailable' because of 'environmental' reasons. What these 'environmental' reasons are, the mechanisms the Department intends to use to prevent mining on account of these reasons, and the location of the leases included in this 'unavailable' category remain a complete mystery at this point." Or, "another justification is that existing leases are not in economic mining units, in other words, not in large contiguous blocks. The Department cautions against the 'fallacy of aggregation' when asked about the vast tonnages under lease. Again, the Department has failed to provide data which would substantiate their claim." In another instance "the Department admits that it has very sketchy information on nonfederal coal, yet speaks with remarkable authority on the need for more leasing."[6]

In short, leasing opponents really wanted the Interior Department to determine in advance of federal leasing precisely how much coal would have to be mined in the next few years and which specific sites would offer the locations best suited for this mining. Leasing should proceed only if it turned out, on making such an analysis, that some of the preferred sites contained unleased federal coal. This expectation was reflected in the view that for the federal coal-leasing program "to work well, relatively complete information on coal nationwide, as well as in the west, would be essential."[7]

Reflecting their commitment to planned-market concepts, the Interior economist-designers of EMARS II disagreed strongly with the view that everything must be centrally planned in advance. They considered that inherent uncertainties made it impossible to achieve such predictive accuracy. Even if it were possible, the cost of assembling all the necessary information and analyzing it would be prohibitive. Moreover, it would be unnecessary. The EMARS II leasing program was designed to provide a market mechanism to achieve virtually the same goals as environmentalists advocated—development of just the right amount of coal and tracts with the lowest overall mining and environmental cost. The central planning implicit in environmentalists' demands required a capability to plan that the federal government certainly lacked at present. Barring a nationalization of the U.S. economy, it might never possess such a capacity. The Director of the Office of Policy Analysis argued that "no better way exists, to our knowledge, of assessing the public interest, than through the procedure of nominations and land use planning we have decided upon. This procedure is far superior to the alternative . . . of central planning in ignorance of the information on values and costs only market forces can provide."[8]

In sum, looking at the cost to society of a small delay for a new draft programmatic EIS on EMARS II, there did not seem to be an urgent need for new leasing and thus the costs probably were not great. But the benefits were not large either. The basic structure of the leasing program was not likely to be changed much, and in any case, changes could later be made apart from the EIS. The Department did not agree with the demands of its critics that it must demonstrate without any doubt that new federal leasing was an absolute necessity. Even with more time, it would resist in principle a request for what it considered to be such a wasteful and ultimately futile effort in central planning.

The debate over the need for more federal coal leasing constituted another important battleground in the conflict between the ideologies of conservationism and planned-market liberalism. Critics of Interior leasing efforts sought to impose conservationist forms of comprehensive coal planning as a precondition for undertaking any new federal leasing—or even adopting any coal program. Plans showing the exact contribution of any new federal coal to western coal development would have to be prepared, proving its urgency. As examined in Part II, proponents of the planned market within the Interior Department sought to resist pressures to undertake such planning, contending that it was infeasible and would be a waste of time and money.

The Anatomy of a Bad Decision

Given the lack of a clear-cut answer to the need for a new draft EIS, a lively debate ensued within the Interior Department. In July 1975 the Secretary of the Interior finally decided to release the final programmatic EIS for EMARS II without a new draft.

Aside from the substantive merits discussed, the Department's decision left its coal program seriously exposed to legal challenge. For one thing, the basic changes in the leasing program from the draft to the final programmatic EIS had prevented public review and comment on most of the main features of the EMARS II program. Probably just as important, the final programmatic EIS lacked any quantitative analysis of the impacts of future coal development in the West. Granted that the Department could not predict with any real accuracy how much coal development would occur, it still would be disingenuous to suggest that the Department had no idea whatsoever. Certainly some broad range—perhaps very broad—could be set. The decision to resume federal coal leasing entailed not only the establishment of a set of policies and procedures for leasing coal, but also, as a practical matter, a commitment to accommodate a level of western coal development that very likely would fall somewhere within this range. Given the intense controversy over the consequences for the West of major new coal development, there was a fairly evident legal necessity—even if little real policy grounds—for the Interior Department to perform an analysis of the environmental impacts of a range of levels of western coal development, including lower and upper bounds.

The Interior Department was not unmindful of these legal concerns at the time, although in the end it simply chose not to heed them. Jack Horton, Interior Assistant Secretary for Land and Water Resources, noted that the programmatic EIS for EMARS II "does not 'quantify' the impacts of the proposed leasing program; it provides only a qualitative analysis." In addition, Assistant Secretary Horton observed that the new program "is different in significant degree from the draft that was first calculated." These and other factors meant that the "risk of litigation" was "substantial." Moreover, there was little to gain from rushing the programmatic EIS anyway because "no leasing will be possible in the Northern Great Plains for 12–18 months because of the Judge Skelly Wright decision in *Sierra* v. *Morton*." In sum, Assistant Secretary Horton recommended issuance of a new draft prior to a final programmatic EIS: "A decision to go 'draft' will reinforce our improving relations with Western Governors, inasmuch as they would clearly have the opportunity to comment on and influence a new coal-leasing program, as opposed to seeing it as already determined in a 'final' EIS. Neither they, nor the public, have had the opportunity to examine or comment on our proposed diligence requirements, commercial quantities or our present EMARS program which is fundamentally different (market-place versus government allocation leasing) than that described in our original statement."[9]

As will be examined in the next chapter, the Interior Department did indeed lose when the expected NRDC legal challenge materialized. The delays in leasing that then resulted were far longer than any Interior Department official in 1975 would have considered possible. Out of a short-sighted impatience to have a federal coal-leasing program in place, a strong desire to take an immediate symbolic step toward greater national energy independence, and a serious misreading of the legal situation, the Interior Department made a major policy mistake.

The Department went ahead and released the final programmatic EIS for EMARS II in September 1975. This release was followed up by some further decisions on program details and the full EMARS II coal-leasing program was formally announced by Interior Secretary Thomas Kleppe in January 1976. The Department requested and received nominations from industry on tracts for new leasing in the summer of 1976. Regional EISs were set in motion, which would analyze new federal leasing proposals, as well as other proposed federal coal actions in each of ten regions. The Department expected to resume federal coal leasing sometime in 1977 or 1978, depending on the speed with which the regional EISs could be completed.

The Federal Coal Leasing Amendments Act of 1976

The Congress was also aware that federal coal in the West had become a central element in national energy policy. Development of federal coal, along with other energy resources, might cause major and often controversial changes in western states. The administration of federal coal was guided by a fifty-year-old law which had received little congressional oversight over the years. These circumstances combined to make it important that Congress take up the subject of federal coal leasing and either bless the existing system or make the modifications it thought were necessary. The acceptability to the West of federal leasing actions and support of western coal development would be greater under a new statute reflecting the current thinking of Congress.

Perhaps more to the point, the western states also had some conditions to be met before they would accept new leasing. They wanted a bigger share of the financial returns from federal coal; they sought assurance of wide public and state government participation in leasing decisions; and they wanted more firm assurance of environmental protection than the administrative mechanisms of EMARS II—which ultimately rested on agency discretion—could provide.

The Federal Coal Leasing Amendments Act, enacted in August 1976 over a veto by President Gerald Ford, served these purposes. The law largely reaffirmed the EMARS II program that the Interior Department had been putting together over the previous two years. Diligent development within ten years was required; continuous operations were mandatory, except in the case of payment of advance royalties; preference-right leasing was abolished; comprehensive land-use planning prior to leasing was mandated; competitive bidding and receipt of fair market value were further important requirements. The act also made some changes, however. It increased the royalty rate for surface mining from 8 percent to a minimum of 12.5 percent, matching the oil and gas minimum royalty. Special lease sales for rural electric cooperatives and other "public bodies" were required. Fifty percent of lease acreage had to be offered with allowance for deferred payment of bonus bids. A government coal exploration program was to be maintained. Provisions were included for a new system of exploration permits with the federal government entitled to information obtained. The government was directed to ensure "maximum economic recovery" of its leased coal. In order to encourage more rapid development of leases, a mining plan had to be submitted within three years. Consultation with state and local governments and the general public was required in the preparation of land-use plans. Finally, and perhaps the most important change, the state share of bonus, rental, and royalty revenues was increased from 37.5 percent to 50 percent.

The new leasing law for federal coal necessitated a number of minor adjustments in the EMARS II program. In some cases changes had already been incorporated in revised regulations by the end of 1976 and in other cases soon would be. The Interior Department was ready to move ahead with a resumption of federal coal leasing, now with a congressional blessing in the form a new coal-leasing act that clearly had been written with such a resumption in mind.

Planning and Politics

In the progressive writings on planning the importance of its separation from politics is regularly emphasized. In some formulations planning activities were even to be carried out by special bodies of planner-technicians supervised by independent boards of citizen appointees. Thus a 1931 Department of Commerce publication typical of such views stated that

> planning should be designed to cover a long period of years, much longer than the term of office of any single city council. Legislation on the other hand is designed to meet pressing and immediate needs. . . . Although the two functions, planning and legislation, are both important, and essential to the efficient working of city government, yet they are quite different from each other and involve different considerations, differing points of view and different talents and interests. The two functions, therefore, need to be reposed in two separate bodies.[10]

Such expectations proved wholly unrealistic under democratic government. Planning was never very much separated from politics; indeed, few people would now advocate that it should be. However, the early planning proponents may well have been correct to doubt that the political process could ever achieve successful long-run planning. Not the least of the problems raised by a blurring of politics and planning is the limited tenure of political leadership in office. How can long-term rational plans be prepared and carried out when the whole cast of key planner/politicians can change every four years—or even more frequently at some state and local levels. New officials coming in may well have much different policy objectives, which they are unlikely to subordinate to the plans of their predecessors. The difficulties raised for orderly planning by wholesale shifts in top leadership were well illustrated in the federal coal program by the consequences of the change from the Ford to the Carter administrations.

9. The Judiciary Mandates Central Planning

The election of Jimmy Carter brought into high-level administration positions a number of previous staff members of environmental organizations. In several cases these individuals had been leading public critics of EMARS II and had strongly opposed any renewal of federal coal leasing. Indeed, an NRDC staff member who had filed suit against the coal programmatic EIS took a key position in the Interior Office of the Solicitor. Not surprisingly, therefore, the Carter administration quickly made known its intentions to thoroughly re-examine federal leasing. The President's Environmental Message in May 1977 announced the objective to achieve a "reform of the Federal coal leasing program." In this regard the most important part of the message was its direction to "the Secretary of the Interior to manage the coal leasing program to assure that it can respond to reasonable production goals by leasing only those areas where mining is environmentally acceptable and compatible with other uses." Further presidential policy guidance specified that the secretary should "scrutinize the existing leases (and applications for preference right leases) to determine whether they show prospects for timely development in an environmentally acceptable manner."[1]

In the summer of 1977 the Interior Department embarked on a full-scale review of the federal coal-leasing program. The Department acknowledged that there existed "substantial pressures to begin production from outstanding leases, and to issue additional leases." However, on the other side "these pressures are resisted by agricultural and environmental interests asserting that present leasing and mine approval practices would allow development of coal in places, and in ways, that would severely damage the natural resources, environment, and local economies of several regions in the West."[2] Cecil Andrus, the new Secretary of the Interior, assigned his Assistant Secretary for Land and Water Resources, Guy Martin, the responsibility to design a new coal-leasing program.

Western Coal for Western Use

As discussed previously, one part of the environmental movement finds the idea of doing damage to the environment morally repugnant. Environmental protection is thus often seen as an ethical question in much the same way that an abortion opponent would consider protection of the fetus or an art collector preservation of a masterpiece as moral questions. This moralistic element of environmentalism clashes with the attitudes of economists and market proponents who see the matter as one of benefits and costs. Instead of a question of right and wrong, economists seek to find the answer to the question: Is the environmental damage great enough to justify the cost of avoiding it? The latter attitude dominated the design of the EMARS II coal-leasing program, but was challenged in the new Carter administration.

Economists considered that western coal development should be promoted as long as the benefits exceeded the cost. The benefits reflected the value of the coal to any customer, wherever he might be. The costs were the direct mining costs including environmental mitigation measures, plus the cost of environmental damages that could not be avoided. Considered in this fashion, the substantial amounts of western coal that forecasts indicated would be consumed in the eastern parts of the United States were no special concern, as long as benefits truly exceeded costs.

Many environmentalists, however, saw large-scale shipments of coal from West to East in a different light. If environmental damage infringed on basic values that were noneconomic in nature, one region of the country should not ask another region to accept environmental damage for its benefit. To do so was morally offensive. Why should the West be degraded for the benefit of the East? Such attitudes in fact are not so unusual; many other social decisions are also removed from the realm of economic transactions. For example, in modern times a rich young man cannot pay a poor young man to take his place in the military draft—even if the latter is perfectly willing. This practice is considered morally offensive and has been prohibited in the United States since the Civil War.

The moral element of the environmental movement was reflected in the coal policy of the new Carter administration. The administration held the view that "overall economic impacts of coal use are most favorable when coal production and development take place near the consuming markets." This meant specifically that "production increases should be encouraged in the eastern and midwest coal regions" for eastern consumption.[3] As one observer remarked, "The new Administration sees federal coal leasing policy as a key tool in shaping the pattern of coal development nationwide," and its efforts were likely to "result in a leasing policy designed to discourage the use of western coal outside the West."[4]

In addition to restraining federal coal leasing, another key element of the administration's coal strategy was to curtail the source of eastern demand for western coal. As noted previously, this demand had arisen because, despite much higher transportation costs, western coal had a low-sulfur content and thus could avoid the necessity to install the very expensive scrubbers that high-sulfur eastern coal generally required. The solution adopted was straightforward: Impose a requirement that scrubbers be employed by all utilities without regard to their ability to reduce emissions by burning low-sulfur coal. The result could be to require the installation of very costly scrubbers to remove sulfur from coal that actually had very little sulfur to begin with—a direct affront to economic rationality if not to the new environmental ethic. With Carter administration backing, a coalition of eastern coal miners and environmentalists opposed to western coal development pushed through a provision in the 1977 Clean Air Act Amendments mandating scrubbing of all coal in new power plants.[5] As the Interior Department reassessed the situation, "The effect of the BACT requirement would be to shift some production from the west to the east, because the BACT requirement was assumed to require scrubbers on locally available coals and hence eliminate the eastern and mid-western markets for low-sulfur coals from distant sources."[6]

One purpose of the Interior Department coal-policy review was to assess the consequences of reduced demand for western coal on federal leasing needs. Other elements of the review included an effort to "develop criteria and process for designating Federal lands" that would be "unsuitable for coal development."[7] Decisions would have to be made on how to handle existing leases and preference-right lease applications in unsuitable areas. Diligence and continuous operation requirements would be reexamined as well as a number of other features of the old EMARS II coal-leasing program.

Jumping ahead of the pending NRDC court suit, Interior Department officials also committed themselves to write a new programmatic environmental impact statement: "The final product of the review plan will be specific proposals for policies and procedures to govern the management of the Federal coal-leasing program. The Secretary will decide whether to approve these proposals only after the Department has completed an environmental impact statement on the proposal."[8] Given the comprehensive nature of the coal-leasing review and the likely time required to prepare a new programmatic EIS, a delay in resuming federal coal leasing of at least two years was virtually assured. Outside observers concluded that "the Energy Minerals Activity Recommendation System (EMARS), the cornerstone of the coal leasing program developed under former Interior Secretary Thomas Kleppe, appears on its way to being junked or substantially modified." As a result, "the magnitude of the changes contemplated almost certainly precludes any full-scale resumption of federal coal leasing before 1979, and could mean there will be no full-scale resumption of leasing until sometime in the 1980's."[9] The latter estimate indeed proved to be a good one.

NRDC v. Hughes

Before the coal-leasing review could get very far along, however, the much-awaited decision was received involving the suit by the Natural Resources Defense Council (NRDC) against the programmatic EIS for the old EMARS II program. An attempt in June 1977 to reach a settlement with NRDC had come close but failed. In September 1977 federal Judge John Pratt of the D.C. District Court harshly criticized the Interior Department and declared the programmatic EIS for coal leasing to be wholly inadequate. Judge Pratt precisely instructed the Department as to its future course of action. Within sixty days it must publish the old programmatic EIS and request public comment. After that the Department was required to issue a new draft supplemental EIS to the old programmatic EIS. Then, a final EIS should be issued and at least thirty days later "the defendant Secretary of the Interior shall personally reevaluate federal coal-leasing policy, based on the information contained in the new final EIS, and shall make a new decision as to whether a new leasing program shall be instituted and, if so, what kind of program it should be."[10] What the judge had in mind in requiring that the secretary must "personally" reevaluate the Department's leasing policy was left unspecified, although leasing was of course carried out under the general authority of the Secretary of the Interior.

As a policy-making instrument the judicial process works slowly—a significant liability in many cases. NRDC had filed suit against the Interior Department in late 1975 and almost two years had passed before the decision was reached. If appeals had been pursued, two more years would probably have gone by. Because of this delay, in 1977 top Interior Department officials found themselves being sued over the environmental impact statement for a coal program of a previous administration that they had already disavowed. Moreover, some of the current top Interior officials had been among the most vocal public critics of the program they were now defending in court.

The political process and the judicial process thus by now had the same general objective—a full reassessment of coal leasing—but were contending for control over how this assessment would be carried out. Under NEPA the judiciary had the upper hand. The judiciary thus established the terms for the policy review.

Given little disagreement about the need for a full-scale review, the most important question was the extent to which the Department would be allowed to continue activities directed toward future federal coal leasing. Judge Pratt opted for a draconian solution, ordering "that federal defendants, their agents and employees, and all those in concert or participation with them, are enjoined from taking any steps whatsoever, directly or indirectly, to implement the new coal leasing program, including calling for nomination of tracts for federal coal leasing and issuing any coal leases, except when the proposed lease is required to maintain an existing mining operation at the present levels of production or is necessary to provide reserves necessary to meet existing contracts and the extent of the proposed lease is not greater than is required to meet these two criteria for more than three years in the future."[11] In short, the order called for shutting everything down related to coal leasing until the NEPA process was exhausted and a new programmatic EIS had been issued.

This order had some curious results. It meant that for the time being environmental studies of proposed new lease sites had to be called off. Studies of the cumulative environmental impact of future new leasing proposals in western regions were judicially enjoined. The judge included a ban on any activities towards processing preference-right lease applications. The Interior Department, however, had an undoubted legal obligation to process these applications and issue the leases if commercially viable coal deposits could be shown. Environmental studies of preference-right lease applications, which could be used to determine whether the Department should seek to buy back the rights to protect the environment, were prohibited. Finally, land-use planning to determine areas where coal leasing might not be suitable in the future was precluded. As matters turned out, these activities—all of them aimed at environmental protection—were almost entirely halted for about two years by Judge Pratt's order.

The Interior Department had had ten regional EISs underway, which included proposed mine plan approvals, new competitive leasing, processing of preference-right lease applications, and coal-related rights-of-way, in one case for a new railroad line. These regional EISs had been initiated in 1976 to set the stage for a resumption of federal leasing in 1977 and 1978. All the parts of the regional EISs pertaining to any new federal leasing were removed from them, in many cases leaving only an empty shell with a mine plan or two. Environmental studies of leasing issues were halted; later on, whole new regional EISs for leasing would have to be prepared.

The result of the judicial order was that if the review of leasing policy and the need for leasing showed that there was an urgent need to resume federal coal leasing, the Interior Department would be almost totally unprepared to undertake it. Then, probably rushing to make up for lost time, the Department might well have to make a more rapid and less careful examination of the environmental and other impacts of federal leasing at particular sites and in broader regions. Again, as matters turned out, this proved not to be an idle concern; preparations for the first few lease sales were hampered by the lack of earlier environmental and economic studies.

Judge Pratt might simply have enjoined the Interior Department from undertaking any actual new federal coal leasing until it had completed the reexamination of its leasing program and decided on the need for such leasing. In effect, however, he had banned any contingency planning—a normal part of the operation of almost any organization—for possible future leasing.

The judge's opinion gave little explanation for the drastic shutdown order. Judge Pratt may possibly have thought the chance that federal leasing would be needed soon was so slim that no contingency planning would even be worth undertaking. Of course, this would seem to belie the Judge's assertion that his concern was only an adequate impact statement; that "the determination as to whether, when, and how to lease more federal coal remains largely within the Secretary's discretion."[12]

Indeed, the most likely possibility was that Judge Pratt considered it his responsibility to rebuke or punish the Interior Department for its moral transgressions in designing and defending its coal program so ineptly and in being so determined to scar the West with what could be marginally necessary coal development. The tone of the judicial opinion was stern and the judge took it upon himself to make the determination that "the Final EIS was not prepared with objective good faith."[13]

The Reasons for EIS Inadequacy

With concern to substantive content, Judge Pratt's basic criticism of the programmatic EIS for EMARS II was "the inadequacy of the Department's explanation of the EMARS program." He was most concerned about the lack of a clear justification for a shift from a central planning orientation to a market orientation. The Interior Department "without the proper explanation" had "substituted a procedure of industry nominations, designed to provide BLM with data as to the location and quantity of coal to be leased, for the original allocation system in which federal agencies related inventoried federal coal resources to national projections of coal-derived energy needs."[14]

The other key criticism was "the Department's insufficient showing that the institution of a national coal leasing program was even necessary." Judge Pratt relied partially on the General Accounting Office's 1976 report which stated that "Interior does not have an adequate basis for determining whether additional coal is needed." He also noted that there was already far more federal coal leased than could be used any time soon, that "even with a significant increase in production, . . . the amount of federal coal currently under lease could satisfy such demand for many decades."[15] No mention was made of the Interior Department rebuttal that unleased federal coal would probably be superior in overall environmental and economic characteristics and that a definitive answer on this matter would be virtually impossible; any attempt to determine exactly how much federal coal was needed would be prohibitively expensive and a misuse of public funds. As the Interior Department had long asserted,

> The existence already under lease of large amounts of Federal coal does not in itself mean that further leasing is not needed. Transportation cost for coal is high, and the constantly-changing geographic pattern of the demand for energy means that a coal deposit that was valueless last year may be a valuable low-cost source of supply next year. If Federal coal which is not yet leased is such a low-cost source, and if we do not stand ready to lease it, National energy costs will be unnecessarily raised, and the public interest will suffer. If on the other hand Federal coal leased long ago no longer has value as a low-cost source, it is not in the public interest that it be produced, and our policies will encourage relinquishment of such leases.[16]

The court argument concerning the need for leasing in fact involved some contradictory elements. Environmental groups claimed, and the court seemed to agree, that there was no need for additional federal coal because so much previously leased federal coal was already available in addition to large amounts of nonfederal coal. But if further federal leasing would actually have no impact on western coal development, it would also have no environmental impact. On the other hand, while there was doubtless enough coal already available in an absolute sense, renewed federal leasing might still have an impact on the location of coal production by making better quality coal available. This could cause some rearrangements of western coal production patterns. Conceivably, if lower-cost federal coal were newly leased, the total level of western coal development might also rise somewhat. Presumably, this was what many environmentalists actually feared, despite their public statements that there was no need for new leasing. But, of course, if this actually happened, it would show that there had really been a need for new leasing.

Moreover, in this case new federal leasing would also very likely be beneficial to the environment. Certainly, newly leased federal coal in the late 1970s or early 1980s would be more carefully examined for adverse environmental consequences than had the coal leased in the 1960s, when there was no systematic environmental examination whatsoever. The main effect of new leasing would probably be to shift production from private coal and old federal leases to newly leased federal coal. As Governor Ed Herschler of Wyoming said in 1978, "I am inclined to favor renewed leasing as a means of assuring optimal market selection among potential lease tracts on the basis of current information. As you know, both the regulatory and economic terms which influence production have changed markedly since existing leases were granted. The optimal economic and environmental pattern of Federal leases is not likely to be the present pattern."[17]

In short, renewed federal coal leasing might be almost wholly innocuous—certainly not a trivial possibility—or if it did have some impacts, this fact in itself would show that the need for leasing had actually existed. Moreover, the new leasing would probably have beneficial environmental consequences.

If there was a serious drawback to renewed federal leasing, it would be to add to the oversupply of already leased coal. Later, it would then be necessary to cancel a larger number of existing leases for failure to be diligently developed. Or, to avoid such cancellation, coal companies might offer bargain coal and rush into premature production ahead of market demands. Issuance of more federal leases might also complicate the problem of predicting where coal development would actually occur, creating a more uncertain environment for state and local land-use planning. These possible problems were not mentioned, however, in Judge Pratt's opinion.

In considering the basis for the judge's opinion, it seems likely that he, like many other participants in the federal leasing controversy, was influenced heavily by a faith in a specific ideology. In *The Closing Circle*, Barry Commoner argued that "on ecological grounds it is obvious that we cannot afford unrestrained growth of power production. Its use must be governed by over-all social needs rather than the private interests of the producers or users of power."[18] Judge Pratt similarly showed real skepticism about the substitution of a market for a government-planned leasing system. As a result of Interior-proposed actions, critically needed planning might lose out: "The policy of long-range environmental planning would be defeated."[19] By enjoining implementation of any actions "whatsoever" to move towards future federal leasing, the judge was striking a firm symbolic blow against evil private interests and in favor of planning for the public interest. Much like his D.C. colleague in the federal courts, J. Skelly Wright, Judge Pratt was showing himself to be a true believer in the conservationist gospel.

Given the flaws in Judge Pratt's ruling, especially in virtually prohibiting any planning for federal coal-lease sales that would occur only after a final programmatic EIS for coal leasing had already been issued, the Interior Department might well have stood a good chance in an appeal. However, the Department decided against appeal. There was little enthusiasm among the new Interior team for a rapid renewal of federal leasing. The more important objective was a comprehensive review of the entire leasing program—something which might take considerable time. If the Interior Department were under court order, it would be easier to deflect industry and other pressures for new federal leasing that were bound to arise. The Department did succeed, however, in negotiating some relief from the court order in an agreement it later reached with NRDC. NRDC itself may have even been taken aback by the all-encompassing sweep of the judicial injuction.

In *NRDC* v. *Hughes*, the judiciary effectively mandated central planning for federal coal leasing; it would be virtually impossible to prepare an EIS meeting the court's requirements without such planning. In doing so, it was faithful to a long conservationist tradition. The Interior Deparment would spend most of the rest of the 1970s attempting to comply with this decision. Central planning would no longer be a vague, if inspiring, ideal left to be translated into reality sometime in the future. Now was the time. It would be a real world test for the central tenets of conservationism: Could government planners genuinely deliver on the scientific rationality promised both in the law and in the theory? Although the conservationist foundations for public land management had been laid seventy-five years earlier, there had been few such real world tests before.

IV. Conservationism Put to the Test

10. A Lesser Role for the Coal Industry

Although the new Carter administration team in the Interior Department, the courts, and outside environmentalist critics all agreed in 1977 on the necessity for a major redirection of federal coal leasing towards conservationism, there was less agreement concerning the exact design for the future coal program. No doubt EMARS II had given too great a role to private interests, especially the coal industry, and much more government planning was needed. But what would a reasonable industry role be? Where would government planners get information which industry had previously been expected to supply? Just what kinds of planning should the government itself now undertake? There were few definite answers to these and other basic questions.

In the summer of 1977 three task forces were set up in the Interior Department to examine coal-leasing issues. The members were drawn to represent the departmental assistant secretaries and the key agencies involved in coal leasing. Although the three task forces proved unwieldy and were soon replaced by a single coal-leasing office, they began a process of sorting issues and options for Department resolution that went on until the final formal adoption of a new leasing program in June 1979. The strategy was first to decide on the basic structural framework for federal coal leasing and then, progressively, to fill in the finer details. The coal review essentially began from scratch—EMARS II was unceremoniously abandoned.

Determining Acceptable Leasing Areas

The first issue formally raised was the old central question of the roles of the market versus government planning and allocation. As a memorandum prepared for a leasing review task force commented,

> There is one overriding issue which will determine much of the shape that the coal leasing program will take. This issue is the role of private industry and market pressures vis-a-vis government planning. Will the government set rates of coal leasing nationally and/or by region? Will the government determine the specific tracts to be developed? To what extent will private industry wishes be given a major role in reaching these decisions? If private desires are to be important, through what channels will they be expressed?[1]

By October 1977 the task force addressing this issue had circulated several drafts and had a final options paper ready for secretarial decision. The options paper addressed the following three key subjects: (1) "the relationship of the Department of the Interior to the private sector at key stages in the planning process"; (2) "whether and in what manner government-developed information or private-sector information is used as the basis for coal leasing decisions"; and (3) "the overall process for arriving at the locations and level of leasing, both nationally and regionally."[2] While these basic questions had been the subject of much past discussion within the Department, the new options paper focused on them more explicitly than had previously been the case and offered the chance for a new set of answers.

The options paper indicated that White House direction had already settled one potential key issue, a central matter in past leasing controversies. The new program would closely tie levels of leasing to DOE goals for coal production. Such an approach would "be a major departure from past coal leasing proposals which were not based on goals for production on a National or regional basis, but where levels were keyed directly to industry expressions of interest."[3]

The remaining fundamental issue was the role of private industry in deciding the areas and specific tracts to be leased, once the total amount of leasing was settled through the production goal process. Three options were identified. The first relied most heavily on industry information, asking industry to designate broader areas in which it had an interest in leasing, and then later to identify specific tracts. This option was regarded as essentially the same as EMARS II and was never seriously considered. The second option would have the government make the first cut on broader areas acceptable for leasing without any special industry input, but then have industry nominate specific tracts which would be given the most attention for leasing. The third option would exclude industry altogether until the point of the lease sale; industry would not enter the picture except to bid on tracts previously selected at government initiative.

The biggest advantage of option one was said to be that it "takes advantage of resource and market information and expertise currently available in industry" and "minimizes requirements for additional government planning resources." On the other hand, the biggest drawback was that "industry nominations prior to setting leasing levels could dominate the subsequent decisions to the detriment of environmental and community values." Option two addressed this concern by initially excluding industry inputs; it "assures that new Federal coal development would be located *only* in the specific areas in which Government believed it to be most desirable." However, "without initial information from industry nominations, Government might unknowingly exclude coal development in areas where there is strong demand." This "might force coal development to less economically and/or environmentally suitable locations." Finally, option three, total exclusion of industry input, "provides for the most complete Government control over coal development" and thus has the "greatest assurance that adverse socioeconomic impacts and environmental impacts will be minimized." However, it "requires the Government to have more extensive company planning information, such as projected coal markets and non-Federal coal holdings." As a result, "higher costs and manpower levels would be required to carry out the program."[4]

In some ways, this set of options represented the elevation of symbol over substance. The presence of a major coal industry and the fact that this industry would actually have to mine the coal could not be denied. Neither could anyone deny that the coal industry had important information on coal types and qualities, coal requirements of utilities, and other key matters. Under any reasonable planning system more information should be better than less. Hence, how could the government even consider not asking for certain highly significant information for coal planning that could be obtained at little expense? To say that receipt of the information from industry would bias the planning process would be a confession of the large gaps in existing government information; it would say that the government had little knowledge of the coal resource and, therefore, receipt of industry information would so predispose matters as to dominate government decisions. But, if government actually knew so little, how could anyone expect it to take over the full responsibility for planning for western coal development?

Seen in a dynamic context, however, the options paper did address a more valid question. Would the government take steps to improve its information base and develop the capacity to match industry information? If so, it could substitute government for industry coal data and other facts and figures. As long as industry information was being used, the pressure for the government to develop its own information capabilities would be minimized. Congress might well refuse money for the government to generate geologic and other data which industry was already capable of supplying. Thus, government might never escape from reliance on industry sources and never acquire the base of knowledge required for the assertion of a strong independent government hand in planning and controlling western coal development. Hence, for the Department supporters of such planning and control, a major issue was indeed at stake.

Interior Secretary Andrus chose option two, under which the Interior Department would identify broad areas in which federal coal leasing would be precluded, and other areas would be identified in which leasing would be permitted. The key, however, was the means by which the areas would be selected—a matter not fully addressed in the options paper. Areas under consideration for leasing would not be subject to a full balancing of their coal development value and environmental cost and then ranked against one another. Such an effort would be very costly, among other liabilities. Rather, certain uniform national standards of unacceptable environmental impact would first be established—in the terms of the coal progam, "unsuitability critera." Using these predetermined criteria, areas with unacceptable impacts would be eliminated from consideration for leasing; and the remaining areas would then be made available for leasing. As described by a later Interior Department task force, "This process is designed first to identify large areas with potential coal reserves which do not conflict with critical environmental concerns (unsuitability criteria) . . . and then to facilitate choice of tracts from among those acceptable areas which meet regional production goals while minimizing environmental and socio-economic costs."[5]

Since there was an abundance of coal in the West, and as long as the standards for unacceptable coal areas were not too broad, it could be assumed that there would be adequate supplies left available of inexpensive and otherwise desirable coal. The Department believed this to be the case: "Based on coal reserve information currently available and reasonable projections of coal production from existing mining plans, the need for additional Federal leasing over the next ten years will likely be small relative to the amount of available Federal coal not under lease."[6]

Within the remaining wide areas determined to be acceptable for leasing, the coal industry would make specific proposals for lease sites, to the point of delineating proposed tract boundaries. The Interior Department would then focus on these nominated tracts in checking for site-specific environmental consequences that might be unacceptable, but had not previously been uncovered in area-wide examinations. Within the designated "acceptable areas" industry desire to lease thus would be a major influence in determining tracts offered. The Interior Department, however, would still have to make a further selection if there were many more tracts available for leasing than its lease sale target. But the potential for mistakes would not be great. Unless there were major differences among tracts in coal development value, not much additional money would be worth putting into tract selection, because the poorer environmental tracts already had been sorted out.

The new leasing system thus offered a pragmatic balance between the need for more information and the high cost of such information. It reflected a recognition that at some point the cost of further information could exceed the benefits in improved patterns of coal development. It would pay to spend large amounts on careful balancing of coal development values and environmental values if there were large differences in these values among coal tracts. However, the government clearly would not want to spend $10 to determine that tract A was $5 cheaper to mine than tract B. Administrative costs and convenience turned out to play a significant role in the design of the leasing system.

In this decision the Interior Department in fact took its first step back from conservationism. A pure application of conservationist principles would have led to selection of option three—comprehensive government planning of all aspects of leasing without industry involvement. Indeed, some members of the new Carter management team strongly advocated this course. But others with major doubts as to the Department's actual ability to identify the appropriate coal to lease without industry help, as well as the high cost of such an effort, pragmatically prevailed.

Concessions to Environmental Symbols

The pragmatic solution achieved was accompanied, however, by a heavy rhetorical overlay. Secretary Andrus was concerned that the new system not appear a defeat for government planning. He also wanted it to be free of any taint of his predecessor's EMARS II program with its use of industry nominations and emphasis on the market. Hence, the secretary emphasized that the Department should "make certain that industry clearly understands that their comments (not nominations) are welcomed and solicited at the same time the public comments on the environment and land use planning." He also wanted it to be clear to industry that "their 'tract interest' can be expressed after we have determined 'acceptable coal areas.'"[7] The specific EMARS II term "nominations" was banished, and would henceforth be replaced by industry "indications of interest."

Politically, the decision to designate unsuitable areas where federal coal leasing would not be considered at all was an important plus. It affirmed for the Department's environmental consitituency the moral principle that there were values automatically higher than coal development value. Like the value of a human life, an endangered species, or a wilderness, other values such as protection of critical wildlife habitat or farming in alluvial valleys were not to be assessed in terms of ordinary economic benefits and costs, but required defense regardless of the costs. The decision of Secretary Andrus to declare areas unsuitable for coal development without first assessing their coal value seemed to significantly expand the range of areas removed from bloodless economic calculation.

If EMARS II had been designed in significant part by William Moffat, his role was taken in the new coal program by Joseph Browder. Starting as an environmental activist in the 1960s, Browder had become a leading strategist for environmental groups in Washington in the early and mid-1970s. He had joined the Carter campaign team and been a principal contributor to its environmental statements. Possibly slated for a White House job, a dispute within the Carter team resulted in his becoming a staff assistant to the Assistant Secretary for Land and Water Resources. Nevertheless, his close White House connections and long background in the environmental community gave him a major influence in coal program design, especially in the first year or two of the Carter administration.

Browder summarized the difference in symbolic emphasis between the old EMARS II coal-leasing program and the new leasing program of the Carter administration as follows:

> Under both systems, site-by-site social/economic/environmental analysis would occur, and balancing trade-offs would be made, prior to leasing a particular tract. But in the first, industry drives the system, with Federal resource managers forced to react to industry initiative, wherever Federal

> coal is found. There will be controversy, and unhappiness from winners and losers from all interests, no matter what system is chosen. The conflict is really between the old system in which Federal resource managers were seen as respondents to industry plans for disposition of public lands and resources, and the more active owner-manager Federal role called for by FLPMA, the Coal Leasing Amendments Act, and the policy statements of this Administration.
>
> When the Federal resource manager drives the system, initial planning decisions are made in response to a balancing of environmental and renewable resource values against the need to achieve certain levels of coal production. The capacity of the Public Lands (and other affected resources) to absorb or recover from damage can be measured against both the need to meet production goals, and the ability to meet that need through distribution of production to alternative areas. National and regional interests in the management of all Federal resources and the exercise of all Federal responsibilities are balanced against each other—not against the economic interest of a particular coal-using company. Company interests are satisfied, at a late stage in the process, when, after a required level of production from Federal reserves is established, and after Federal resource managers determine where that production could take place without unacceptable damage to other resources, appropriate amounts of Federal coal are offered for lease.[8]

The central theme is the great importance of fully retaining the basic decisions in the hands of the federal government, while holding industry influence to a minimum. To some extent this theme reflected an awareness, as described in chapter 6, that full protection of environmental and noncoal-use values required that the federal government make the final decisions. It was also a reflex reaction to memories of the long years in which ranchers and private industry dominated public land decisions—a symbolic repudiation of these years. Holding industry at a distance was also a symbolic affirmation of the conservationist commitment to government planning and its aversion to the profit motive of the market as a primary influence in allocating natural resources.

Finally, government control was a means of retaining control over the rate of western coal development, which environmentalists feared market forces might drive too rapidly, if they were not already doing so. Instead, the new leasing system should take control over the rate of western coal production. This pragmatic concern was voiced explicity on occasion:

> If the Department wants to keep prospective leasing somewhat in line with prospective production, it's better to avoid pressure to over-lease. We want to avoid the position of choosing between tracts, each of which is socially/environmentally acceptable, each of which has been offered for bid, but one of which has to be rejected because we decide that development of the other tract will produce enough coal to satisfy DOE's production goal. Whatever

our production limitation may be, it's better applied early than late in the leasing process.

A leasing system that has Interior, not industry, decide how much coal should be offered for lease is complementary to a system that has Interior, not industry, make the initial choice of appropriate areas from which to produce Federal coal. It is difficult to imagine a workable system that could accommodate industry initiative in one element and Interior in the other—whether we're considering tonnage or geography.[9]

The secretarial option paper on the basic framework for federal coal leasing was the first of a large number to be eventually prepared in the development of the new coal program, although most would deal with lesser matters. Government option papers sought to lay out all the alternatives and typically listed their pros and cons. They were partly an effort by government analysts to assist and, in some cases, also to pressure top decision makers to examine systematically all the implications of their decisions. In this case the option paper may have had such an effect. Given the strong conservationist rhetoric and philosophy of the new Carter administration, it might otherwise have ignored the pragmatic concerns that led it to select a mixed industry-government—rather than an all-government—system of planning for federal coal leasing.

The desire to keep the coal industry at a distance, shown by the Interior Department during the early Carter administration, illustrated tensions that had surfaced long ago in the original conservation movement. Its frequent hostility toward big business stemmed partly from a populist element in conservationism—a dislike of any organization too big or too powerful. Yet, as discussed in chapter 1, the central tenets of conservationism were scientific management, comprehensive planning, and more efficient use of natural resources—and it was precisely the large modern corporation that was most likely to promote such objectives. As historian Samuel Hays has observed, "The conservation movement did not involve a reaction against large-scale corporate business, but, in fact, shared its views in a mutual revulsion against unrestrained competition and undirected economic development. Both groups placed a premium on large-scale capital organization, technology, and industry-wide cooperation and planning to abolish the uncertainty and waste of competitive resource use."[10]

But at the same time, as Hays notes with respect to President Theodore Roosevelt, he conceived "of the good society as a classless society, composed, not of organized social groups, but of individuals bound together by personal relationships. . . . He viewed the fundamentals of social organization as personal moral qualities of honesty, integrity, frugality, loyalty, and 'plain dealing between man and man.' Thus, he was 'predisposed to interpret economic and political problems in terms of moral principles.'" Reflecting such views, Roosevelt was hostile to the trends of urbanization and industrialization. He considered that "the good society was agrarian. He would have opposed bitterly any effort to turn back the industrial clock, yet his ultimate scheme of values was firmly rooted in an agrarian social order." As a result, Roosevelt really was of conflicting minds. "One, a faith which looked to the future, accepted wholeheartedly the basic elements of the new technology. The other, essentially backward-looking, longed for the simple agrarian Arcadia which, if it ever existed, could never be revived."[11]

This conflict has never really been resolved in American society. Early Carter appointees perceived members of the coal industry as organizations which could exert undue pressures on the Interior Department. Moreover, the very efficiency of the industry in producing coal threatened to override older social values in the West. Indeed, the rural, sparsely populated West was one of the last remaining American refuges from the spreading urbanization and industrialization that had distressed Roosevelt. On the other hand, the coal industry possessed the necessary technical planning skills and knowledge to make a success of western coal development. It represented the embodiment of scientific management in coal production. Moreover, social harmony—to say nothing of political survival—required that a reasonable relationship be established with the coal industry, along with other important interest groups. Hence, after starting off in an adversary relationship with the coal industry, the later Carter years would be marked by efforts to move towards greater coordination and cooperation.

11. Central Planning in Operation in the United States

The Role of Coal Production Goals

The EMARS II aim to adopt a planned-market approach to determining the level of federal coal leasing had been much questioned from the beginning. Indeed, as examined in part II, the actual market plans devised were seriously flawed. In its 1976 review of the coal program the General Accounting Office had advocated that federal leasing be undertaken instead to achieve government-set production goals.

> We believe that some fundamental attempts should be made . . . to relate the amount of Federal coal required to meet national goals to any program of renewed leasing. At this stage, we do not know how much coal is expected from the Federal lands, how much is already under lease that can meet such an expectation or how much more, if any, should be leased.
>
> Specifically, we believe that the Secretary of the Interior should more precisely identify what role should be played by Federal coal resources in meeting national coal production goals.[1]

Environmental critics of Interior coal leasing offered similar views. Commenting critically on the EMARS II program, NRDC stated that "The Department must explain . . . the conclusion of the President's national energy plan that there is a lack of demand for coal, rather than a lack of supply. . . . There is, of course, no way for the public to determine whether the demand for coal will be more or less than these projected production figures unless the Department supplies figures estimating future demand."[2]

As noted in the previous chapter, the President's Environmental Message of May 1977 adopted such views as official government policy. Production goals and leasing targets were to be worked out by the Departments of Energy and Interior. A Memorandum of Understanding for coordination between the two Departments in developing production goals was later signed. The memorandum sought to specify procedures for setting the goals and their use in federal energy leasing. Coal production goals would be part of a much wider effort in which the "development of an integrated national energy policy by the Department of Energy requires the coordinated treatment of Federal resources as a constituent part of national energy planning."[3]

Production goals based on comprehensive national energy planning thus were to be the driving force in the new coal-leasing system; they would ensure that federal coal development was undertaken only in the amounts that national energy needs required. Guy Martin, Assistant Secretary for Land and Water

Resources, considered that "this is the central question in the design of a Federal leasing system—the issue of what triggers the need to transfer coal from the public domain to the private economy. By directing that the leasing system respond to production needs the President has overcome the problem which caused so much conflict and confusion in the development of the previous Administration's coal policy." Hence, "when the appropriate contribution of Federal coal to the nation's energy budget over any given time period is identified, the Department of the Interior will lease enough coal to respond to those production goals."[4]

However, the proponents of such a system based on production goals had little familiarity with the details of implementing this approach. Production goals are common enough to centrally planned socialistic economies. But in the United States, what would a national production goal for coal mean in the absence of any central economic plan? How far in the future should the goal be? How often should it be revised? These questions were still unresolved—indeed hardly even considered—well after the decision had been reached to make production goals a centerpiece of the new federal coal program.

Defining a Production Goal

By early 1978 the Interior Department had made several key decisions. First, a new Office of Coal Leasing Planning and Coordination would be created under the Assistant Secretary for Land and Water Resources, to take charge of the coal-leasing effort. Assuming an actual need for such leasing had been found, Interior Secretary Andrus had announced his intention to resume leasing under the new permanent program by October 1980—just in time to beat the November presidential election. The Department had also decided to prepare a brand new draft and final programmatic EIS for coal leasing, essentially washing its hands of the old programmatic EIS for EMARS II. The commitment to be ready to lease by late 1980 put the development of a new leasing program on a very tight schedule, especially since the programmatic EIS and then further new regional lease sale EISs would have to be completed sequentially, as a result of Judge Pratt's order. Thus, the Department determined to have the main features of the new coal program worked out by the summer of 1978. This program would provide the proposed action for the programmatic EIS to be released in final form by the spring of 1979.

With the decision made to base leasing levels on production goals, implementation was turned over to the working level technicians—mostly economists—in the Departments of Interior and Energy. The Office of Leasing Policy Development in DOE had the responsibility for transmitting coal production goals to the Interior Department. The Interior Assistant Secretary for Policy, Budget, and Administration was assigned responsibility for Interior's coordination with DOE, a task mostly carried out by the Office of Policy Analysis.

Because higher-level Interior and White House policy makers really had little idea how coal production goals might actually be calculated, the economic and energy modeling technicians in the Interior and Energy Departments received very little guidance as to the procedures to be followed. They were left to shift for themselves, but to do it quickly, because there was little time for long-range contemplation.

There were not in fact many options. DOE had previously sponsored construction of models for making forecasts of future coal production. In 1978 it was in the midst of improving earlier forecasting models and also of building large new ones, not only for future coal production but throughout the energy sector. The only realistic possibility for future "production goals" would have to be the DOE forecasts of actual future coal production.

The forecasts of future coal production would at least reflect current policies affecting coal production. These policies included pollution control requirements, nuclear safety restrictions, oil price decontrol, utility coal-burning requirements, and a number of others which could significantly change production expectations for coal. What basis was there for changing future coal production goals, other than shifts in such policies, which would actually cause coal production levels to change? It would make little sense to define a goal as simply a hoped-for level of coal production, but for which there was no specific method of implementation or particular reason to believe it could be achieved. In the United States a goal could not have the same significance as in a socialist economy, because there was no direct control over the private sector which would actually have to implement the goal.

Nevertheless, the use of a production forecast significantly violated the spirit in which the production goal approach had been adopted. The use of goals was intended to limit the influence of industry in determining the level of leasing. But in using forecasts, the role of industry might simply be sanitized; all the forecast would amount to would be an attempt to predict industry behavior. Production levels in accord with industry objectives would be derived by means of a large DOE computer model, rather than accepting the same information more directly from industry inputs.

The question of whether leasing should be driven by direct industry requests or government goals took on major symbolic significance. As a technical matter, however, the whole issue was really miscast. If national economic efficiency were accepted as the basic objective, the purpose of federal coal leasing should be to assist in minimizing the cost of producing coal and other energy for the whole nation. The DOE forecasting model worked by means of a great number of computer calculations to minimize total national production costs for given levels of utility power generation and other coal-use demands. But in a competitive coal market the same coal production levels would result if private companies each sought to produce their coal at the lowest prices and to supply it at the lowest cost. It is a fundamental tenet of modern economics that the market works to achieve national economic efficiency—the same objective of DOE forecasting. The DOE modeling did not incorporate environmental costs, except insofar as they directly raised the cost of coal mining or precluded use of certain coal altogether. As a result, except to the extent that imperfect information or irrational behavior led to mistakes by the government or private companies—admittedly no trivial matter—the regional distribution and overall levels of coal production would be the same under either approach. The real question was whether in practice government goal setting or the market mechanism was likely to come closest to the efficiency objective they shared.

The First Production Goals

In June 1978 DOE supplied its first coal production forecasts to the Interior Department, prepared for the years 1985 and 1990. The forecasts were broken down by region, coal type, surface and underground mining, and other categories. Comparable sets of coal production forecasts, revised to reflect changing events and assumptions, were also supplied in April 1979 and December 1980.[5] Such regional forecasts have served as the actual production goals in the procedures established for setting future levels of federal coal leasing.

There are major uncertainties in forecasting future coal production. To some extent DOE acknowledged this uncertainty in its decision to supply not one forecast, but three—a high, medium, and low forecast. Rather than one goal, there would actually be three, raising again the question of what a "goal" really meant. In the initial forecasts supplied by DOE, the high forecast for national coal production in 1990 turned out to exceed the low forecast by 67 percent.

Mistakes in forecasts could be of two kinds. First, the forecast of total national coal production could be wrong. Second, even if this forecast were accurate, the regional breakdown might be off. The economic consulting firm which had developed the DOE coal forecasting model came close to disavowing its use at the level of regional breakdown employed by DOE. The company cautioned that "it must be understood that no model is accurate at its lowest level of regional disaggregation. . . . The fundamental data upon which the supply curves are based . . . do not warrant undue confidence in highly disaggregated estimates." As a result, "The regionally disaggregated production estimates should be used as a guide to how much production would occur in a region. Undue credence should not be given to any single estimate. The more aggregate numbers . . . should be viewed as being much more accurate than the disaggregated numbers (i.e., Powder River Basin in Wyoming)."[6]

An illustration of such problems promptly arose in the Uinta-Southwestern Utah coal region. For the key Utah part of the region, the model projected considerably less production than was already planned by coal companies. To eliminate this obvious discrepancy, DOE simply substituted the planned production level for the computer model result. As a consequence, however, a new anomaly appeared—an almost identical high, medium, and low forecast for both 1985 and 1990, varying only within 26 and 28 million tons per year.

Successive DOE projections for the San Juan River coal region bounced over a wide range. By the end of 1980 DOE had provided three different sets of coal production forecasts for 1985 and 1990 for this region. The first, which was made in 1978, forecast coal production to be 22.8 million tons in 1985 (medium assumptions); in 1979 the medium forecast was lowered all the way to 13.4 million tons; but then in 1980 it was raised back up to 35.4 million tons. The medium forecast for 1990 had even more extreme fluctuations—from 58.4 million tons in the 1978 forecast to 16.8 million tons in 1979, and then back up to 57.6 million in the forecast made in 1980.

The most important coal region is the Powder River region. Here, as well, the coal production forecasts proved very erratic. In 1978 the medium forecast for 1990 was 396 million tons. In 1979 the same medium forecast was raised to 418 million tons, but in 1980 it fell all the way to 294 million tons. Such rapid shifts reflected the great sensitivity of coal production forecasts for individual regions to changes in demand and supply assumptions.

Forecasts of total coal production for the nation have also shown considerable variation. The 1978 medium forecast for national coal production in 1985 was 1,116 million tons. By 1979 this forecast had already fallen to 1,033 million tons, but then it rose back in 1980 to 1,118 million tons—about the 1978 level. The 1990 medium forecast for the nation showed a similar pattern—first falling 4 percent and then rising 11 percent in successive years.

The difficulties experienced by DOE are not due to lack of effort or failure to employ the best modeling techniques available. They simply reflect the natural hazards of forecasting in an extremely uncertain environment. Consider just the major unexpected changes that occurred in the two years after DOE's first coal production forecasts were provided to the Interior Department:

—the OPEC price shock of 1979
—the Three Mile Island incident, casting more doubts on the future of nuclear power
—decontrol of domestic oil and natural gas
—the natural gas scarcity turning into potential abundance
—a sharp dropoff in projected demand for electric power
—new legislation authorizing large government subsidies for a synthetics fuel industry
—substantial increases in expected demand for coal exports.

Several examples illustrate the inherent hazards of forecasting future coal production. Following the fourfold OPEC price boost of 1974, the nominal price of oil held steady and the real price had even declined somewhat by 1978. Hence, the world oil price assumptions used in the 1978 coal production forecast were widely criticized as being too high. Responding to such complaints, DOE lowered some of the assumed levels of world oil prices—the high oil price for 1990, for example, was reduced from $30 a barrel to $23.50 a barrel. But then only a few months after DOE had lowered oil price assumptions, world oil prices skyrocketed again. More than doubling, they exceeded even the high level which DOE had previously been assuming would not be reached until 1990.

In making forecasts of coal exports, little consideration was given in 1978 to the possibility that high world oil prices and the unreliability of Middle East oil would cause a large shift to steam coal, especially in Europe. The potential for large increases in U.S. coal exports to meet such demands was similarly overlooked. For the 1978 forecast, the levels of coal exports in 1990 were 76.3, 77.2, and 77.1 million tons for low, medium, and high forecasts respectively. The negligible spread from the low to high forecasts certainly did not represent the genuine range of possible export levels. Exports were simply set at the high end of recent historic experience. By 1980, however, there was a growing public awareness of the potential for greatly increased coal exports. The 1980 forecast raised the coal export levels for 1990 to 159.9, 178.2, and 258.6 million tons.

A similar story can be told with respect to synthetics. In 1978 there was considerable optimism about synthetics production, and coal used in 1990 for synthetics was forecast to be 56.2 million tons (medium forecast). By 1979 the mood had become much more skeptical, and the 1990 forecast was slashed in half to 28 million tons. But then OPEC promptly turned things around, and in 1980 the synthetics forecast jumped sevenfold to 198.3 million tons. (Although as of this writing matters have again turned around, and the 1980 forecast now looks much too high.)

If it had not been for the increase forecast in 1980 for future export and synthetics production, the level of national coal production predicted for 1990 would have been significantly lower. The DOE forecast for both electric utility and industrial coal use in 1990 fell sharply between 1978 and 1980.

Erratic behavior in coal production forecasts was caused by a period of rapidly unfolding events and major uncertainties. It has also reflected a tendency to understate genuine uncertainty, stemming in part from an unwillingness of forecasters to acknowledge, even to themselves, the uncertainty of the future. Forecasters are, after all, selling a product—the ability to predict the future—and they are not disposed to cast strong doubts on their own product. Public pressures also push forecasters to promise greater certainty than they can actually deliver. Consumers of forecasts commonly prefer not to acknowledge the true risks they face and seek comfort in an undue faith in forecasts. This human trait is nothing new; astrologers, palm readers, oracles, medicine men, and other seers have appealed to it throughout history.[7]

The Department of Energy spent many millions of dollars on its energy modeling. It employed graduates of the finest universities trained in economics, statistics, and computer science. Yet, in all likelihood, equally accurate forecasts could have been made by a single individual very knowledgeable in the coal industry, using a hand calculator and with a few days of background inquiries and data gathering. Admittedly, to some extent DOE's problems were caused by the great rush in which the first coal forecasts were prepared. Future forecasts will probably show greater stability from year to year and be significantly improved in other respects. But it is unlikely that they will ever achieve very exact predictive skills; indeed, serious doubts must now be acknowledged as to the possibility of "scientific" computer forecasts ever doing much better than well-informed personal judgment supplied with some basic data.

The setting of production goals and the energy planning of the Department of Energy is probably as close as the United States has come to a full-scale central planning operation. The deficiencies of the DOE forecast were an illustration on a limited scale of the technical complexities of calculation and coordination that have bedeviled central planning efforts for whole national economies—some of which have considerably less gross output than the U.S. energy sector. At one point, the Interior Office of Policy Analysis even toyed with the idea of hiring a U.S. specialist in Soviet economic planning to look at the federal leasing program. But the idea was turned down as politically too risky. The same political sensitivities led the Interior Department to change the length of the initially proposed "five-year plan" for federal coal leasing to a "four-year plan."

Even if the coal forecasts had proven much more stable, the United States still would have been a very long way away from adopting true central planning. The "goals" would not really be goals in a controlling sense, because U.S. coal production was carried out in the private sector. Although the federal government could make coal available by leasing it, it could not guarantee that the coal would be produced. U.S. coal production would be determined by the sum total of financial incentives, regulatory policies, and other influences on the coal industry. By forecasting private production under such influences, the government could assess whether it liked the forecast result. If other coal output levels were desired instead, government would have to alter its leasing policies, create new industry incentives, or change coal regulatory policies. Forecasting—not goal setting—thus was what was really called for. In short, the actual coal program shift to production forecasts, even if still formally in the name of production goals, was a pragmatic substitution consistent with the real world of the U.S. private economic system.

12. Setting Coal-Leasing Targets

The DOE production goals for western coal were the starting point in the new coal program procedures to determine planning targets for the leasing of federal coal. But the first question was whether any new federal leasing was needed at all (whether the target perhaps should be zero). Indeed, this question had been the central controversy of coal leasing in the 1970s. Opponents of western coal development sought to discourage it through a variety of strategies, but most of all through preventing resumption of federal coal leasing. Given the predominance of federal coal in the West, if federal leasing could somehow be shut off, western coal development might gradually slow to a crawl.

The opposition had been waged by contesting the scientific adequacy of the official explanations given for resuming federal leasing. Judge Pratt in *NRDC* v. *Hughes* had spoken for many people who were offended by the EMARS II rejection of conservationism, and who were not prepared or inclined to ask whether basic conservationist tenets could be wrong. Judge Pratt had thus ordered that the new programmatic EIS for federal coal leasing must contain a full analysis of the consequences of not undertaking any leasing. In light of his decision, and with its tone very critical of the Department, the ability to put in place any permanent leasing program appeared to depend upon a better demonstration that there was a significant, genuine need to lease more federal coal right away.

The method of assessing leasing need was straightforward. For each region, an estimate would be formed of current coal company plans and other commitments for future coal production—coming from already issued federal leases as well as nonfederal coal. If these current commitments exceeded the production forecasts (goals) for the region, no leasing would be needed. However, if these commitments were less than the forecasts, creating a potential shortfall, a federal leasing target would be set in amounts appropriate to the expected federal share of new regional coal production.

As noted previously, public doubts about the need for new federal coal leasing had been fueled by the huge overhang of federal coal reserves in already issued leases. The basic question was: Why should federal coal leasing resume when there was already far more federal coal under lease, in total, than could be used for many years, although not necessarily in each region?

The New Assessment of the Need for Leasing

The results of estimates for 1990 are shown below in Table 12.1, which is taken from the assessment of the need for leasing presented in the 1979 coal programmatic EIS. In the Powder River region, for example, the total committed (planned and likely) production for 1990 was 226.1 million tons. This was much below the medium production forecast for 1990 of 396.1 million tons. Several other regions showed similar shortfalls of committed production below the production forecast, indicating a need for new leasing. Based on such estimates and similar estimates for 1985, the Interior Department drew several conclusions.

> An assessment of the need for new Federal leasing based on projections of demand and supply levels thus does not produce an unambiguous picture. For 1985, there appears to be little need for new leasing, except in one region, the Green River–Hams Fork Coal Region. For 1990, there could be some, but probably not a large, need for new leasing to reach low projected production levels. On the other hand, achievement of medium and high 1990 production levels would require extensive development of new sources of western coal production, especially in the Powder River, Green River–Hams Fork, and San Juan River Coal Regions. Because more than 70 percent of the coal in the six western Federal coal states is owned by the Federal Government, new federal leasing would make a major contribution in achieving such development.[1]

The Interior Department was concerned, however, that too much might be made of the numbers in table 12.1; it offered several cautions concerning such analyses. For one thing, the possibility of major forecasting errors was acknowledged: "Forecasts of future energy demands and supplies are subject to many uncertainties. The uncertainties increase the further in the future the forecast is made." Errors could also exist on the supply side because a certain amount of currently planned production might not ever materialize: "Not only the uncertainty surrounding future levels of demand, but also the uncertainty of any given tract passing through the steps from potential tract to fully operational mine must be taken into account in assessing leasing needs." As a result of these uncertainties, the Interior Department warned, as follows: "The absolute need for new leasing to meet national energy objectives thus depends on which assumptions about future energy demands and the role of western coal in supplying those demands prove to be most accurate. Uncertainty also exists about planned production estimates. . . . Since it is impossible to know at this time which assumptions and estimates are actually correct, government policy must be flexible."[2]

In fact, the actual record of estimates of coal company plans for future production showed a high level of variability. Different DOE estimates of 1985 planned production in the Fort Union region varied from 24 million tons to 87 million tons.[3] In another region, Uinta–Southwestern Utah, DOE estimates varied from 20 to 45 million tons. In the key Powder River region, the variation was from 128 to 280 million tons. Such variations essentially reflected differences in the certainty that planned production would actually ever be realized. The lowest estimates examined production plans that had a very high certainty

Table 12.1. Summary of planned, potential, and projected production, 1990 (million tons)

Coal region	Total 1985 planned production	Likely production from existing leases without mine plans	Total planned and likely production	Production potential PRLA surface reserves	Total production potential	1990 DOE projections		
						Low	Medium	High
Fort Union	21.8	[a]	21.8 [b]	[a]	41.5 [c]	21.9	20.6	34.5
Powder River	219.1	7.0	226.1	48.5	274.6	173.7	396.1	602.9
Green River-Hams Fork	49.8	6.8	56.6	0.3	56.9	105.9	149.5	177.7
Uinta-Southwestern Utah	47.2	23.3	70.5	1.8	72.3	25.1	28.3	27.9
San Juan River	24.0	8.5	32.5	11.3	43.8	34.5	58.4	72.5
Denver-Raton Mesa	3.0	[a]	3.0 [b]	[a]	23.6 [c]	5.4	6.8	6.6
Totals	364.9	57.3 [d]	422.2 [d]	90.5	512.7	366.5	659.7	922.1

a. Not disclosed because of confidentiality requirements.
b. Does not include likely production.
c. Total includes likely production and PRLA surface production potential.
d. Total includes Fort Union and Denver-Raton Mesa regions.
Source: Bureau of Land Management, U.S. Department of the Interior, *Final Environmental Statement: Federal Coal Management Program* (Washington, D.C.: Government Printing Office, April 1979), pp. 2-52.

—generally for which contracts had already been signed. The highest estimates may have included some plans that were barely more than scraps of paper, tentative proposals still at the talking stage, but which companies for one reason or another had chosen to label as a plan.

Although the Interior estimates focused on planned coal production, the more relevant figure would have been future capacity. However, the development of capacity estimates would have been even more difficult, and this was judged not to be a feasible option.

Currently planned production thus was really used as a surrogate, only because it was the only available number that could be defended. It was deficient in that it would show future production capacity only to the extent that coal companies had already projected future coal demand to exist and had gone ahead with plans to meet this demand. In short, currently planned production is no more than the sum of individual coal company forecasts of their own share of future coal production. Rather than a true estimate of supply capacity, it is really an alternative form of production forecast. If coal companies have reasonably accurate knowledge of each other's production plans and thus successfully coordinate to avoid two companies planning to meet the same future demand, the total amount of currently planned production may well serve as a better forecast of future production than forecasts derived from elaborate computer modeling such as employed by DOE. The difference between a computer-derived forecast of future coal production and the sum of current planned production of coal companies hence may show nothing more than a disagreement between two methods of forecasting future coal production.

Where future coal production is so far off that coal companies have not yet planned for it, then computer forecasts of production at that time will inevitably exceed current company plans. This is the primary reason why planned production by coal companies for 1990 fell significantly short of the computer-based forecast for 1990 production. Instead of the Interior Department's portrayal of this shortfall as showing the need for leasing, a more accurate statement would have been that it simply showed that companies do not plan production ten years ahead.

Hence, despite a surface reasonableness, the Interior assessment of leasing need actually did not show much about the true need for leasing. The Interior Department had refused to provide such an assessment in the earlier EMARS II program on the grounds that it would mean little and not be worth doing. But by 1978 there was no choice in the matter; however logical or illogical, valid or invalid, the past refusal of the Interior Department to develop production goals and compare them with estimates of already committed production potential had become a powerful symbol of Interior intransigence and unwillingness to accommodate a wide range of public opinion. Indeed, top Interior officials, for the most part, subscribed to the general public view.

The Interior assessment of leasing need has a close parallel in an analysis long performed by the Forest Service—an earlier manifestation of conservationism. The basic Forest Service timber assessment has consisted of a forecast of future timber requirements which is then compared with timber production likely to occur under current harvesting plans. Based on such assessments, Gifford Pinchot and many other professional foresters regularly predicted the imminent arrival of a "timber famine"—an event which, of course, never happened. More recently, instead of the famine myth, there have been less dire predictions of timber "shortages," which have been used to justify requests for more investment funding for the Forest Service.

Outside economists reviewing Forest Service assessments have been critical of their neglect of basic economic principles of supply and demand. Under normal market workings, future shortfalls in supply below demand will stimulate price increases that cause demand to fall and supply to rise. In the actual event, supply must, of course, be equal to demand. Instead of the prospect of a timber shortage, more valid questions are: What adjustments will occur to bring supply and demand into balance? What costs will these adjustments bring? Are there better ways to balance supply and demand? In a book examining past Forest Service timber planning, *The Depletion Myth*, Professor Sherry Olson comments that the Forest Service "completely ignored the complex factors that constitute economic supply and economic demand and the relationship between them. The economic facts of supply, for example, were very different from the physical facts of supply. The physical supply of timber was diminishing, as everyone could see, but the economic supply probably was not."[4]

The Real Consequences of Not Leasing

Just as in the case of federal timber, the real question for a federal coal-leasing assessment was the impact on supply and demand of the further availability, or lack of such, of federal coal. Coal supply would always equal demand in the end; recognizing the role of prices, and other adjustment mechanisms, production could not fall short of demand. The real questions were: How would the pattern of coal production differ with and without further federal coal availability? What would be the benefits and costs of the adjustments required to maintain a balance of coal supply and demand? But these were less dramatic and more difficult to explain than a looming "famine" or "shortage," either of coal or timber.

If there were no further federal coal leasing, several general types of adjustments to the absence of new federal coal supplies would occur. Within a given coal region, new mines would be opened or old ones expanded, using nonfederal coal or federal coal already available in existing federal leases. If much of the coal in the region were federal, leaving few nonfederal alternatives, new mines would probably be economically and environmentally inferior to potential mines precluded by the lack of federal leasing. Interregional shifts in coal production would also occur, from coal regions more dependent on federal coal to those less dependent—in general, shifting production from the West with its high percentage of federal coal to the East with its mostly private coal. A third type of adjustment would be a rise in the price of coal, more in some regions than in others, but generally reducing the total national demand for coal. Either less energy would then be used, or other energy forms such as oil and gas would be substituted for coal.

The valid question in assessing the need for federal coal leasing is not the ability to reach production goals, but the cost of the adjustments required without it—especially the extent to which coal production costs and coal prices would rise if new federal coal could not be acquired. The Interior Department was not wholly unmindful of this point, although it did not emphasize it to the degree that it might have. It acknowledged the following in its programmatic EIS: "The fact that currently planned and likely production, together with production potential from PRLA's, is not sufficient to reach medium and high 1990 DOE production projections does not mean that these projected levels could not be attained without new Federal coal leasing. . . . There are large amounts of Indian and other non-Federal coal reserves in western regions sufficient to meet almost any conceivable 1990 production requirements."[5]

The Department also undertook a study which sought a rough estimate of the costs of the various adjustments that a withholding of further federal coal would necessitate. The results were reported as follows:

> In order to assess the impact of no further leasing of Federal coal, a special computer study was made in which future western coal development was limited to non-Federal coal and coal in already issued Federal leases. In addition, non-Federal coal dependent on unleased Federal coal for its development was considered unavailable for future mining.
>
> According to the study, the greatest impact of no further Federal leasing would be experienced in the Powder River Coal Region in 1990. Under medium assumptions, production in this region in 1990 is projected to decline by 27 percent if there is no further Federal leasing. The Wyoming portion of the Green River–Hams Fork Coal Region showed a projected decline of 54 percent under a no leasing policy. Other western regions were either not greatly affected or showed production increases due to displacement of coal production from the Powder River and Green River–Hams Fork Coal Regions to these regions. Nationally, coal production in 1990 was projected to decline by 4 percent under a no leasing policy. For 1985, the study concluded that a no leasing policy would cause only minor impacts nationally and within the West.
>
> National oil and gas consumption was projected to rise in 1990 by 300,000 barrels per day if there were no further Federal leasing (medium assumptions). According to the study, utilities would experience on average an eight percent national increase in delivered coal prices. This would cause a 1.7 percent average national rise in electric utility rates. The estimated total

> resource cost to the Nation in 1990 of no further Federal leasing was projected to be $800 million per year.
>
> The regions most adversely affected by a no leasing policy would be in the West, reflecting the fact that western coal supplies primarily western markets. According to study projections, the Rocky Mountain, West North Central, and Pacific regions would experience increases in delivered coal prices in 1990 of 29, 17, and 27 percent, respectively, if there were no further Federal leasing (medium assumptions). These coal price increases would cause overall electric power rates to rise by 6.4 percent in the Rocky Mountain region, 5.9 percent in the West North Central region and 1 percent in the Pacific region.
>
> The principal consequences of leasing less Federal coal than is needed to meet national energy objectives would likely be to alter patterns of coal development, both at national and regional levels. At least on the basis of computer projections, it appears improbable that total national coal production would be greatly reduced.[6]

Although there was very little confidence in the specific numbers given, these projections were perhaps useful in indicating likely directions of change. The description provided might best be considered a metaphor in scientific imagery; the reporting of precise computer calculations heightens and enlivens the presentation of what is an essentially qualitative message. The general attitude in the Interior Department was probably best captured in the following overview assessment included in the programmatic EIS:

> Forecasts of energy consumption and of available energy sources are based on assumptions which are subject to change. Discovery of additional or alternative energy sources, advances in technology, successes in energy conservation programs, variations in the rate of growth of electric power use, and many other factors could cause coal demand forecasts to be significantly revised, up or down. Sound long run government policy must acknowledge this uncertainty, and not assume that today's forecasts must inflexibly govern resource production decisions of the future.[7]

The final step in a valid assessment of the need for leasing would have compared the adjustment costs of the kind described above with the net environmental benefits (or possibly damages) also resulting from no further leasing. Would shifting coal production from the Powder River Basin to other western regions less dependent on federal coal or to the East have environmental consequences beneficial enough to justify the loss of access to low-cost mines in this region? Would shifting coal production within a given western region from federal to nonfederal sites have a desirable environmental consequence? Finally, would producing somewhat less coal nationally and using somewhat more energy from other sources (nuclear power, foreign oil, hydropower) have a beneficial environmental impact? The Interior Department also made some very crude estimates of the environmental impacts of the coal production redistributions resulting from a continued unavailability of new federal coal leases. However, the Department did not draw any specific conclusions as to whether the net environmental consequences would be beneficial overall, to say nothing of whether any possible benefits might actually be worth the cost.

Shifting coal production from the West towards the East would have a mixture of environmental consequences. Reestablishing the vegetation of reclaimed areas is easier in the East than the West, due to heavier rainfall. On the other hand, the lands in the East are more valuable for farming and other uses which would be lost during mining. In many cases, these lands have greater scenic value as well. Surface mining in the East must cover a much greater number of acres per ton of coal, due to the much thinner seams; eastern mining also frequently occurs in steeper terrain where it may cause greater erosion and more conspicuous visual scars. More mining in the East would shift some mining underground, with various environmental consequences. One of the more important would be a greater safety hazard to miners; in contrast to the East, the typical western surface mine is little more dangerous than an ordinary road grading operation.

In most areas the East would already have built up cities and towns with infrastructure and public service capacity in place. New coal mining might be a boon to long-depressed Appalachian communities—some of the poorest sections of the United States. In the West new coal development would mostly take place in sparsely populated areas where it might strain existing public and private facilities. Particular localities might double or triple in population in a few years. These remote western locations have had clean air and water, and an environment free of many of the problems of urban life. As such, besides the attraction to the residents themselves, the rural West has a powerful symbolic attraction for harried urbanites throughout the United States—it might even be seen as an "endangered" type of human environment.

In addition to the factors described, the Interior Department also considered that new federal coal leasing was needed to assure adequate competition in the western coal industry. This argument was supported by the Antitrust Division of the Justice Department, which reported that "resumption of the Federal leasing program with all deliberate speed will have beneficial competitive effects."[8]

A final consideration, as important as any of the factors mentioned above, was the impact of diligent development requirements. One major change from 1975, when the EMARS II program had been proposed, was that the diligent development clock had finally started running in 1976. Old leases issued in the 1960s were now supposed to produce within ten years—by 1986—or be subject to cancellation. Hence, any new mines opened after 1986 could not rely on federal coal unless some new federal leasing occurred. The alternative would be to relax the diligent development requirement, but critics of the Interior Department had long pressed instead for tighter enforcement of this requirement.

Strict enforcement of the diligent development requirement would wipe out the huge backlog of reserves in nonproducing federal leases by 1986. Elimination of this backlog would undercut the primary argument against renewed federal leasing. By 1978 the earliest that federal leasing could resume would be 1980, only six years distant from 1986, surely not too early to begin feeding in new federal leases that would be needed in 1987 and later years. Strict enforcement of diligent development requirements thus would virtually necessitate that new leasing go forward. This was an ironic if often unrealized consequence for the many Interior Department critics who strongly opposed both renewed federal leasing and relaxation of strict diligent development requirements.

Most recently, this argument has in a sense been turned on its head. Wide skepticism about the Interior Department's actual ability to lease much coal has created strong pressures to relax the diligence standard. In addition, serious legal doubts have been raised as to Interior's authority to require diligent development by 1986 for leases issued prior to 1976. The argument is made that Interior cannot unilaterally change the provisions of leases until the twenty-year initial term of the lease runs out. Under regulations issued in 1982, most old leases will have until the mid-1990s to get into production. Partly as a result, the Reagan administration has moved to abandon the 1986 requirement for commencement of production.

Public Skepticism Concerning the Interior Department Analysis

Publication in December 1978 of the Interior Department's assessment of the need for new leasing provoked wide public comment, much of it critical. As examined above, there were major failings in the Interior Department's assessment of leasing need. But these failings were mostly in the basic approach—something many of the critics failed to recognize. Their attacks thus tended to address what were actually only the symptoms of the real problem.

The coal industry showed the greatest awareness that the basic approach of the assessment was faulty. The National Coal Association regarded the attempt to set production goals and match them against planned coal production as offering a specious and misleading precision:

> The impression of accuracy and reliability conveyed by the DES description of the regional production target setting process is seriously misleading. Coal market forces are subject to significant fluctuations over relatively brief periods of time. Long term trends are difficult to predict with accuracy. Even relatively recent data regarding potential production from existing federal leases for which mine plans have been submitted or are known to be in preparation is now of uncertain reliability.
>
> Estimates of production from other federal leases for which mine plans have not been prepared are even more speculative. Demand-based modelling such as is presently used within both the government and the private sector is of particularly questionable reliability, due to the magnitude of the effects thereon of such variables as supplies from foreign energy sources, future contributions of competing domestic energy resources, and the past and

future effects of major changes in the governmental controls on the development, production, transportation and consumption of coal.

The nature of the coal industry, the nature of its markets and the extraordinary variabilities to which it is subject produces a demand-supply situation of such complexity that even the most sophisticated modeling techniques cannot approximate future realities.[9]

Other industry commenters made similar points. The amount of information required for the government to make a conclusive demonstration of the need for leasing would be enormous. The government would have to know the relative economics of potential coal mines, the coal requirements of utilities and other coal users, the ability of these users to substitute among coal and other energy sources (or to conserve on energy), transportation costs from various potential mine sites to all kinds of coal users, and the impact of sulfur content, and other differences in coal quality on the production costs of users. This is far from exhausting the list. Not only would these matters need to be known for the immediate present, but projections of them all far into the future would be necessary. In short, as an executive of a leading coal company stated, "Unless the federal government (by accident or design) *nationalizes* the coal industry, the federal government cannot (nor can anyone else) competently quantify the 'need for leasing.'" There were simply too many unknown and unpredictable factors:

The decisions individual businesses make are usually based on economic considerations known at the time the decisions must be made and are largely unpredictable much in advance of the time when the needs are known and the plans of the individual business dictate a decision be made. (For example, Will there be coal slurry pipelines someday? When? To "where" and from "where"? The "from where" has much to do with the *quality* of the coal. The "to where" could significantly affect the "delivered price" by offering a significant alternative to the rapidly escalating rail rates. Will there be technological "break-throughs" favorably affecting the economics of what today's conventional wisdom says are the economic parameters for studies, forecasts, and projections, or the viable choices available to the market place when the individual producer/user decisions are made by them tomorrow?)

There are many laws, rules and regulations impacting the individual business decisions to be made by coal producers. No computer model or program can predict in a meaningful way the many decisions the market place will make which still determines whether specific producers will go in to the coal business, or open new mines, or shut down or expand existing mines. (For example, OSM has recently reported that strict and immediate enforcement by OSM of just one of the new regulations dealing with non-conforming structures will result in the loss of annual production of 107 million tons of coal.)

No computer model or program can predict . . . whether specific users will use coal at all, or discontinue or expand their use of coal. (For example, new plant locations, and expansions of existing plants, are highly dependent upon

EPA/State air regulations. Where, when and how large will future Class I Areas be? What areas will become the "non-attainment" areas tomorrow or ten or twenty years from now? The plans of users and potential users has already been significantly affected by the changes which the Clean Air Act Amendments of 1977 made in the Clean Air Act of 1970.)[10]

Environmental critics of the Interior Department were no happier. Despite the much greater effort in the new programmatic EIS, environmentalists contended that the Interior Department still had failed to prove the need for new leasing. The assumptions made in the DOE forecasting model were disputed, although not in a very evenhanded fashion. Virtually every assumption favoring higher forecasts of western coal production was strongly challenged by environmentalists, and, in each case, revised assumptions favoring lowering such forecasts instead proposed. In some instances, this tactic gave the right answer. NRDC, along with a number of other environmental groups, criticized the "unrealistic assumptions about the rate of growth in electricity demand." As a result, "the DOE production projections are based on exaggerated estimates of demand." Two years later, in 1980, DOE had significantly lowered its estimates of electricity demand. NRDC also foresaw later DOE corrections in suggesting that industrial demand for coal was probably much overstated. On the other hand, NRDC showed that its crystal ball could be flawed in asserting that "the DOE estimates of coal required for synfuel production and exports . . . are similarly overstated."[11] Two years later projections of synfuel and export use of coal had risen substantially. (Although predictions turned downward again a year or two later.)

Despite the environmentalists' tendency to find only reasons why coal production forecasts for the West were too high, the criticism served effectively to show the lack of a clear objective basis for coal production forecasts, and the importance of subjective views, whether those of forecasters in the government or of someone else. NRDC also criticized the Interior Department for suggesting that cancellation of leases under diligent development requirements would necessitate new leasing to make up for federal reserves no longer available. Instead, NRDC suggested that the Secretary of the Interior make use of the provision in the coal-leasing regulations for a five-year extension of the period in which development must be achieved. But NRDC then promptly showed mixed feelings about this idea: "This is not to argue that such a policy is desirable. Indeed, we recommend that the Department move as quickly and aggressively as possible to determine whether existing leases can and should be developed." At the bottom line, however, given a choice between alternative evils of new leasing or relaxing diligence requirements, NRDC reluctantly opted for the latter: "In the event, however, that coal demand does not grow as quickly as anticipated, it may make sense in many cases to extend the diligence period rather than to cancel an existing lease and subsequently issue a new lease."[12] This was a curious position to take, especially for an environmental organization, because new leases would undergo a process of environmental review far more rigorous than had the old leases which NRDC seemed to favor. It provides another example of the wide impact of symbolic considerations in federal leasing debates. By 1979 renewed federal leasing had such undesirable overtones that it would be opposed, whether it was good for the environment or not.

As discussed previously in the context of the *Sierra* v. *Morton* court case concerning the Northern Great Plains, the theory of comprehensive planning, carried to its full logical implications, suggests that everything must be taken into account all at one time. There is no good way to set limited bounds of a comprehensive plan, because there will always be important factors that would have to be left outside those bounds. Another example of this problem arose in the debate over the need for renewed federal leasing. How could the Interior Department even consider leasing any additional federal coal until the more fundamental question of whether there was an actual national need to burn any more coal had been answered? Perhaps oil and gas, nuclear power, hydropower, less conventional energy sources, or especially energy conservation, would be preferable to greater use of coal. If this were the case, then the Interior Department should not be encouraging coal use by offering new federal coal leases. In short, any development of federal coal should only be proposed in the context of a more comprehensive plan for national energy production. The assessment of federal leasing needs thus could not be confined within even the broad bounds of national coal production issues.

Although this logic might be technically correct, common sense dictated that the Interior Department planning for federal coal leasing should not attempt to recreate a national energy plan. Nevertheless, NRDC did not shrink from such a prospect.

> As we pointed out in our comments on the draft programmatic statement issued by the Bureau of Land Management in 1974, the statement must contain a thorough consideration of "the *major*, national energy alternatives to the proposed . . . program." We also suggested in our earlier comments that it may be preferable to base the comparison of alternative energy options on a careful consideration of the tradeoffs between different energy sources in a programmatic environmental impact statement for the National Energy Plan. In the absence of such a statement, it is incumbent upon the Department of Interior to provide as thorough an analysis of the alternative energy options as possible in order to develop a rational Federal coal leasing program which will best suit the needs of the nation.
>
> Rather than performing such an analysis, the Department of Interior appears to have accepted with little thought the assumption that expanded use of coal represents the main avenue for achieving the goal established in the National Energy Plan of reducing projected increases in imports of foreign oil.[13]

It is doubtful that even NRDC believed in planning and the necessity of a scientifically rational demonstration to this extent. The real aim, as in other instances, was very likely to demand something as a condition for resumed federal leasing that Interior was incapable of delivering.

Other environmental groups offered similar criticisms. As Friends of the Earth complained, "There are several severe problems with the coal supply and demand projections developed for the Draft ES by the Department of Energy. The model inherently overstates both energy demand and supply. The National Coal Model used by DOE also appears to significantly overstate the demand for Western coal as a portion of total coal demand. In addition, the particular iteration of the model used for the draft ES contains unrealistic assumptions which skew the projections towards even higher coal demand. These three characteristics of the DOE projection effort inevitably lead to forecasts of higher Western coal demand than are credible."[14]

Perhaps more surprisingly, Friends of the Earth was in considerable agreement with the coal industry about the slender reed provided by modeling projections: "There is an aphorism in the forecasting business that while there are no magic numbers, numbers are magic. They give a false sense of precision and provide the decision maker with an easy handle on difficult issues. All documentary presentations of energy forecasts are hedged about with qualifying statements which indicate that the forecasts are basically worthless. . . . Needless to say, these hortatory cautions which are always trotted out when projections are attacked, do not prevent the affected decision makers from relying on the numbers in question. If attacked on the validity of the projections, they say 'Well, we said right there in the document that it wasn't perfect,' rather than correcting the errors. Then they go right along using the wrong numbers."[15]

However, Friends of the Earth was unwilling to carry this analysis further to examine the proper role of long-run forecasts in the leasing system, and to ask the obvious question of whether a leasing decision should actually be based on such uncertain forecasts. Indeed, to the contrary, it even faulted the Interior Department for having raised other more valid considerations, such as the environmental and economic effects of shifting patterns of coal development resulting from an absence of new federal leasing. "Not only is the Department's analysis of coal demand and supply inadequate . . . but it attempts to justify the adoption of the program on the basis of undocumented and speculative 'benefits' which have nothing to do with the determination of need for new competitive leasing. Judge Pratt was very clear in asking '*whether* the proposed policy is even *necessary*' based upon reserve, demand, and production statistics, *not* on factors such as competition, administrative convenience, or presumed better patterns of development."[16]

Friends of the Earth found that the analysis which the Interior Department had presented concerning patterns of coal development was inadequate. The Department had not tried to show what the precise impact of a no-leasing policy would be on patterns of development within a given coal region; it had simply argued that there were a priori reasons for believing that such patterns would be improved.

A precise determination within a given region would have required site-specific projections and detailed comparisons of individual coal mine sites, with and without further federal coal availability. This was a potentially very costly and time-consuming undertaking and one subject to major uncertainties. Yet, Friends of the Earth seemed to be calling for just such an analytical effort: "The department also proposes that new leasing is necessary to promote desirable patterns of coal development. This argument assumes that private coal development patterns will be undesirable, a premise unsupported by any evidence in the DES. . . . The analysis . . . fails to compare the preferred alternative with other alternatives, including maximum private land development, to confirm whether it will, in fact, lead to more desirable coal development patterns"[17]

Some Unspoken Purposes of Interior's Need Assessment

The attacks made on the Interior Department assessment of leasing need followed the past pattern of environmental criticism. But this time the Interior Department had tried much harder than before to rationally justify its proposed action, even though still failing by the strict standards of scientific rationality that were applied by environmentalists. The Department had used many people and substantial funds; reams of computer printouts had spewed forth. Top computer modeling experts had been sought as consultants, jointly with the Department of Energy. A leading consulting firm, the Mitre Corporation, was brought on for various forms of technical assistance. Although their accuracy was extremely questionable, many thousands of detailed projections of energy and environmental impacts of western coal development had been made. The Department had finally done what its critics had previously asked; it had made forecasts of future western coal production and compared them with plans for western production in the absence of new federal leasing. Estimates had been made of the precise interregional impacts on U.S. coal production of new federal leasing and even a dollar figure given of the total cost to the nation of not leasing.

If much of this was really offering the fiction rather than the substance of scientific rationality, despite Interior's best efforts, it nevertheless left the Interior Department in a much better legal position. Previously, environmental critics could point to the fact that the Department had refused to undertake any comprehensive planning for coal leasing. It had not made any detailed forecasts, had not examined all the impacts, had not looked at all the alternatives, nor had it performed other tasks expected of a truly comprehensive planning effort. A judge or other observer could plainly see that this was the case. But now the Interior Department had at least tried to perform all of them; new legal challenges would have to address the actual skill and competence with which the tasks had been carried out. This assessment would be difficult for a judge or other layman, since it would require considerable familiarity with computers and scientific terminology. Even for professional economists and statisticians, a good sense might be formed quickly, but to demonstrate definitively the adequacy of the scientific analysis would require examining hundreds of assumptions and the detailed procedures followed in extensive computer calculations. Such an examination would be both time-consuming and expensive; it would probably necessitate hiring expensive professional talent. Few if any potential litigants had the means for such a challenge.

It is doubtful that anyone in the Interior Department consciously worked out these tactics in advance. Nor was the Interior Department especially cynical; it was actually somewhat laggard in following what had already become standard operating procedure. Federal agencies all over Washington in the mid-to-late 1970s were adopting basically the same strategy. In order to try to stem the flood of judicial and other intervention in their internal decision making, volumes of elaborate computer forecasts and other apparently scientific calculations were being prepared to accompany and justify agency decisions. Uncounted numbers of thick EIS documents and other technicial studies poured forth as evidence offered up to the true "scientific" quality of agency decisions.

If a legal challenger had managed to demonstrate clearly the weaknesses in the Interior need for leasing and other central planning efforts, the obvious next question would have been what to do now. The Interior Department had probably performed up to the existing forecasting and planning state-of-the-art —at least within any reasonable budgetary and time allowances. If the results were still very unsatisfactory, and still could not really prove definitively the need for leasing, should the government simply wait for better data and central planning capabilities—with no assurance of their ever being available? That would hardly seem reasonable.

In short, although there was wide criticism of the Interior assessment of leasing need, in the larger picture this criticism might have been beside the point. Whatever its substantive merits, the Interior Department had performed the analysis which the courts and public had demanded. It had accomplished its most important purpose: removing the question of the need for leasing from the jurisdiction of the courts.

Moreover, a new legal challenge to renewed federal coal leasing was not an especially attractive prospect for an environmental organization such as NRDC, which was one of the few that might have been able to mount it effectively. The national pressure for energy development was growing, and public frustration with environmental delays was becoming more evident. Congress might very well have overridden the courts if they had sent the Interior Department back to write a third coal programmatic EIS (or fourth EIS counting the draft for EMARS I). Finally, the new coal program designed by the Interior Department, whatever its deficiencies, contained significant changes to accommodate earlier environmental criticisms, including new criteria for declaring areas unsuitable for coal development. The program had been put together in significant part under the leadership of former environmental critics now working in top Interior Department positions. The Interior assessment of leasing needs thus passed on the bottom line; it paved the legal way for a resumption of new federal coal leasing.

Setting the First Leasing Targets

The draft programmatic EIS for federal coal leasing was released in December 1978. Comments were received on the draft over the next several months and the final coal programmatic EIS was published in April 1979. After allowing 30 days for comment, by early June Secretary Andrus was ready to make his final decisions on the specific contents of the new federal coal-leasing program. Although it was not required, he also decided at this time to schedule tentatively four initial lease sales and to set preliminary leasing targets for three of those sales. This effort provided the first opportunity to see how the system for setting leasing targets would work for an actual sale.

The coal production forecasts (goals) developed by the Department of Energy included not only federal but also private and state coal. Moreover, the DOE forecasts were established for five, ten, and fifteen years in the future. As a result, DOE production forecasts generally did not translate directly into targets for current federal coal leasing; indeed, setting leasing targets required a number of key assumptions and policy determinations. Depending on these matters, the same coal production forecasts could produce quite different conclusions concerning the amounts of federal coal that should be leased.

One key assumption was the amount of time that should be allowed from the lease sale until actual production would begin. The Interior Department estimated that four to seven years would be required to prepare a mine plan, gain necessary permits, sign a contract, and get into operation. The coal industry complained, however, that a longer period was often necessary, especially considering normal permit delays, the amount of time needed to negotiate coal contracts, the time to assemble a full package of coal property rights, and other necessary tasks. In June 1979 Secretary Andrus chose to allow six years in two regions and nine years in a third, but did not elaborate on the reasons.

In setting tentative leasing targets, the Interior Department also had to consider what was likely to happen in situations where projected coal demands for 1986 were not enough to absorb already planned mine output. Assuming the total demand estimates were correct, would most currently planned mines still start up, but produce at levels less than presently planned, leaving substantial excess capacity available for production increases in later years? Or, would some currently planned mines be abandoned altogether, and the remaining mines operate at or near the output now planned, leaving little excess capacity? The answer could make a big difference for federal leasing requirements. The year 1986 was critical because it marked the deadline for getting existing federal leases into production under diligent development requirements (assuming then that they would be enforced). Any mine plans involving federal leases that were abandoned for lack of demand could not be revived after 1986 without new leasing. Secretary Andrus chose a conservative assumption—that excess capacity would not be left in 1986; as a consequence, new federal leasing would have to be undertaken to accommodate all of the federal share of coal production needs for new mines opening after 1986.

Another question was whether to base leasing targets on the low, medium, or high DOE production forecasts. As a secretarial option paper noted, "A determination on this issue reflects an assessment of the costs of leasing too little Federal coal versus too much Federal coal in 1981. Mistakes of leasing too much coal could be corrected by later reductions in the number of Federal coal-lease sales. The adjustments to insufficient 1981 leasing would largely take the form of production shifts from lower-cost Federal to higher-cost nonfederal sources, both within a region and interregionally."[18] On this issue Secretary Andrus split the difference and chose to follow the medium production goals.

A further similar question was the extent to which the leasing target should be increased to allow for the many general uncertainties in its calculation. The Interior Department policy was that it would "not necessarily choose a leasing target that represents an amount of Federal coal just sufficient to reach the regional production goal. Instead the Secretary may choose to adjust the targets upward in order to provide greater assurance of meeting production goals." In adopting this policy it was fair "to say that we regard the consequences of leasing too little as a more serious threat than the consequences of leasing too much."[19] But the Department would move cautiously because it also believed that

> the risks in leasing too much Federal coal are significant, although less obvious than those of leasing too little. . . . One concern is simply the administrative costs to the government in delineating, analyzing, ranking and offering for sale a substantially larger number of tracts than are needed. Excessive leasing would create greater uncertainty about the amount and location of future coal development in the West, making much more difficult the efforts of State and local governments to properly plan to mitigate the expected social and economic disruptions. The ability of the Federal Government to influence coal development patterns to minimize those disruptions would be reduced. Finally, in the presence of the 10-year diligence requirements, excessive leasing could well be disruptive to coal markets. Some companies might enter into premature production just to keep leases and there could be significant administrative costs in cancelling leases that had not been diligently developed.[20]

The Department considered that "there may be other reasons to raise the targets as well. More leasing may be needed to strengthen competition in the coal industry in a particular region—the Antitrust Division of the Justice Department has been particularly concerned about this."[21]

The federal government might want to stimulate competition not only among coal companies but among western states. In the absence of congressional restrictions, the limiting factor on high state severance taxes is the competitive threat that coal production will shift to other states. Montana's 30 percent severance tax might have to be reduced if it drove enough production to Wyoming. But such interstate competition requires that sufficient amounts of federal coal be available in each state to facilitate shifts in coal production. Otherwise, the federal government might find itself, in effect, the cartel manager enforcing production allocations for an OPEC-like grouping of western states with very high severance taxes.

Taking account of all these factors, it could very well be desirable to increase the leasing target much above the amount believed necessary to accommodate coal production forecasts. Secretary Andrus took a limited step in this direction by deciding on a 25 percent increase above the initial target calculation for the Powder River Region.

Outside critics, including several other federal agencies, have pressed for much larger adjustments in setting the leasing target. In Congressional testimony the Antitrust Division of the Justice Department reported its assessment that production goals "will necessarily be imprecise and subject to considerable error." In addition, "even if the coal model accurately determines production needs, leasing only that amount of coal may very well foster competitive problems in the Western coal markets." Another factor was that "the Secretary [should] consider not only the quantity, but also the type, of coal that should be leased."[22] The Antitrust Division recommended that: "Collectively, these problems are quite serious and pose a formidable threat to a national policy of increased reliance on coal. They can be substantially mitigated, however, by leasing two or three times as much coal as will be indicated by the Department of Energy production targets."[23] The Department of Energy, General Accounting Office, and the Council on Wage and Price Stability offered similar views. In 1980 DOE specifically recommended a much higher margin for error ("security factor") in the first lease sale in the Green River-Hams Fork region: "DOE favors a high (more conservative) security factor because (a) industry has indicated that the most favorable coal acreage in the GR-HF Region is not being offered, (b) the cost of over leasing is less than the cost of under leasing and (c) there is considerable uncertainty regarding synfuel and export demands."[24]

In addition to the Green River–Hams Fork target of 531 million tons for an early 1981 lease sale, in June 1979 Secretary Andrus also announced targets of 109 million tons for a later 1981 lease sale in the Uinta–Southwestern Utah Region and 776 million tons for a 1982 lease sale in the Powder River region. A fourth lease sale without a target was announced for 1981 in Alabama.

A Political Decision within Technically Reasonable Bounds

The original coal program conception was a highly deterministic process in which a number of calculations starting from DOE production goals and working backwards would lead to a clear-cut leasing target for each federal coal sale. The necessary calculations would be economic and technical in nature with little subjective judgment. The leasing target thus would be an objective number—the product of a scientifically rational procedure. However, setting leasing targets actually turned out to leave very wide administrative discretion; a broad range of targets could be given a creditable justification by the Interior Department. The most critical question was how much leeway to allow for all the unforeseen ways in which coal production forecasting and federal coal target setting would no doubt go wrong. Formal analysis and expert methods could not contribute all that much to this question.

Nevertheless, the production forecasts and other elements of the leasing target process do play a legitimate, indeed critical, role in getting the leasing decision into the right ballpark. Should the Interior Department be looking in a given region at leasing 50 million tons, 5 billion tons, or something in between? What are the approximate consequences of leasing too little or too much? Without such basic analysis, political leadership would be at a loss in trying to set leasing targets. But once the broad bounds of the decision are set—the ballpark is laid out—subjective judgment must play a much greater role, and political involvement is inevitable.

Although never explicitly acknowledged, it seemed apparent that Secretary Andrus had, to some extent, worked backwards from his preferred leasing levels to the assumptions that would produce these leasing levels. Politically, he seemed to want to come out favoring the encouragement of western coal development and to support the federal leasing levels necessary to sustain further development. But he also did not want to overdo it; federal leasing should proceed in an aggressive but nevertheless still cautious—"responsible"—fashion. His selections of leasing targets were thus significantly affected by the way he felt others would perceive them. The actual amounts chosen fit his political requirements, allowing him to achieve his desired stance. Some people in the Interior Department were in fact surprised that the targets announced were as large as they were. A couple of years later the targets did not look very big, but it was easy to forget that even in 1979 the resumption of any federal coal leasing at all could not be assumed.

It is likely that the technical analysis actually necessary to set the stage for political leasing decisions could have been done with a small fraction of the effort put into it. The size of past efforts can be seen in two lights. On the one hand, much of it simply provided a scientific dressing for decisions that really are not amenable to rigorously scientific methods, but which existing political and legal institutions nevertheless insisted should be made this way. On the other hand, there are some unintended benefits from the necessity to make complex computer calculations and perform other technical studies. These requirements take considerable time and thus impose an orderly decision-making procedure. The way must be paved gradually for the final decision reached; if a decision must come forth from an elaborate technical procedure, radical changes in direction at the last moment become much harder to make. A process thus is imposed which limits the ability of any one individual or few persons to dominate it. Leasing decisions consequently emerge to reflect the middle of the road; there is less chance that eccentric views will be dictated at the last moment by ill-informed top decision makers. On the other hand, if the consensus is wrong, there are few chances that an especially insightful top decision maker, proceeding on a more intuitive basis, will be able to impose his superior judgment.

13. The Advance and Retreat of Land-Use Planning

The idea of a painful tradeoff was antithetical to conservationism, and the biggest reason for its clash with economics. A further example of the denial of the necessity of tradeoffs arises in the area of land-use planning. As discussed previously, the importance of planning has long been a main theme of conservationism, and reflected since World War II in a growing commitment to the use of formal methods of professional land-use planning. Yet, at the same time conservationism has retained its longstanding commitment to "multiple-use" management on the public lands. The problem is that formal land-use planning and multiple-use management really are, to a large extent, opposite and incompatible procedures.[1]

Multiple-use management involves the refusal to commit land to any specific use in advance. It is sometimes contrasted with "dominant-use" management under which timber, recreation, wilderness, or some other specific use is designated as the basic use in an area. By contrast, the essence of multiple-use management is the preservation of wide administrative discretion. Decisions are made case-by-case as specific projects or use proposals arise. By focusing information and management attention on specific proposals with clearly identifiable consequences, multiple-use management is often a very practical mode of operation.

On the other hand, the essence of formal land-use planning is to make as many decisions as possible in advance—to prepare a blueprint of the future uses of the land. Because decisions are made in advance before specific project proposals are received, they must be made on an areawide or zonal basis. The land-use plan shows that timber harvesting is to be permitted in one zone, another area is reserved for intensive recreation, a third area will be kept undeveloped in a wilderness state, and so forth. In effect, formal land-use planning means the adoption of a zoning system for the public lands.

There are many issues raised by such a decision. Is it desirable to have a system in which decisions are left open for wide administrative discretion, or a system based instead on use decisions fixed in advance in the determination of zoning? There is a large literature on just this subject in the field of urban zoning.[2] But conservationists never really faced the issue. Congress has done no better; it enacted legislation mandating simultaneously a philosophy of multiple-use management and the adoption everywhere of formal land-use plans. In practice, the Forest Service and BLM have followed multiple-use management on some lands and have adopted specific plans designating use zones such as wilderness for others. However, over large areas management decisions have been left to be made in the future as an exercise of administrative discretion; it was not considered desirable to designate the areas for any specific uses at the present time. In rationalizing their decisions, the public land agencies have interpreted "multiple use" or "land-use planning" in whatever fashion was convenient for them at the moment, ultimately leaving these terms without any real meaning in agency use.[3] As a result, there has been much criticism on this point, with one observer lamenting that in following multiple-use policies "tragically . . . the complaints on both sides, in fact, are completely justified and the final compromises reflect, rather than harmony, an appalling lack of leadership. The truth is that the Forest Service has no policy. It charts no national course. It simply blows where the political storms blow it, riding the middle of the wind, heading for no port, bent only on somehow keeping afloat."[4]

As in other matters, environmental opponents of western coal development were quick to recognize and seize upon a large gap between conservationist principles and practice. Despite the asserted BLM commitment to land-use planning, in most areas the BLM was actually following the more traditional philosophy of wide administrative discretion under multiple-use management. It strongly resisted having to prepare plans showing certain zones where coal development would automatically be acceptable and other zones where it would automatically be excluded. Hence, BLM's frequent assertion that it employed comprehensive land-use planning could not withstand much scrutiny; when environmentalists began to investigate BLM planning closely, it was soon revealed to be an empty shell. Existing "plans" tended actually to be compilations of motherhood statements with few real world consequences—much like most of their urban counterparts. Criticism of BLM land-use planning became one of the staples of environmental attacks on federal coal leasing. Coal leasing should not resume until a credible land-use planning system was installed.

The pattern thereby set in motion was similar to the need-for-leasing controversy. BLM and the Interior Department would inevitably have to respond to environmentalist criticisms by trying to put conservationist principles into practice. When environmental critics entered government with the Carter administration, they had to live with their earlier insistent demands for land-use planning.

The Development of Unsuitability Criteria

The development of a zoning system for federal coal lands was not characterized explicitly as such. Rather, it was portrayed as the development of "unsuitability criteria" by which coal mining would be permitted or excluded in any given area. By means of this semantic distinction, the various connotations of "zoning" arising from a long history of urban controversy were avoided. But whatever knowledge and understanding these controversies had shed on broader issues of zoning was also lost. In fact, few people in the Interior Department perceived that an "unsuitability criterion" was simply a zoning classification in another name.

The commitment to exclude coal leasing in certain areas actually was first made in the EMARS II program, which provided that leasing would generally occur "where the BLM land use plan . . . has indicated the areas as suitable for mining." Unsuitable areas would include "areas which cannot be rehabilitated or which offer other, overriding resource values."[5]

The Interior Department had assigned the responsibility for unsuitability determinations to the BLM land-use planning system. But BLM had found it difficult to break away from its multiple-use traditions. Accustomed to wide field discretion, BLM never issued any clear guidance as to how the final decision to give one land-use priority over another might actually be made. BLM hoped to avoid a zoning system and to make the unsuitability determinations case by case as specific coal tracts were nominated. Environmental groups, however, were very skeptical of the protection that would be provided by a system of wide field discretion. They much preferred the clear establishment of priorities in advance and the high national visibility of these priorities that a zoning system offered:

> Basically, MFP's [the BLM land-use plans] are nothing more than a rather simple inventory of resources for a given area, with major emphasis on identifying the economically attractive (usually strippable) coal resource. The "public participation" aspect of the program is primarily a series of lectures by Bureau of Land Management personnel, and the written materials and maps—which are the substance of the MFP—have been nearly impossible for the public to obtain so that their participation could be effective. Predictably, MFP's have always concluded that the strippable coal areas are "suitable for leasing," regardless of surface owner desires, environmental conditions or other factors. In at least one instance, BLM asked the coal industry to identify all areas of interest to them, so that the MFP would reflect that interest as a priority.
>
> We conclude, because of the MFP process, the entire EMARS process has a fatal flaw. The assumption that nearly all economically attractive federal coal is "suitable" for leasing, coupled with the completely inadequate mining regulations, results in a proposed program not demonstrably better than the scandalous giveaway program of the pre-1971 period.[6]

Attempting to respond to such criticism and to provide some more concrete guidance, in October 1976 the Interior Department published regulations providing for designating certain areas as unsuitable for coal leasing and specified three standards for such designations: (1) the lands could not be reclaimed, (2) coal mining would pose a risk to public health and safety, and/or (3) coal mining would preclude a higher value use of the coal. Nevertheless, even though some further elaboration was provided, the standards were still very general and clearly left wide administrative discretion.

Aside from the concerns of environmentalists that local values differed from national values, wide local discretion would also create other major problems. There might be wide variation from one local area to another in the type of lands on which coal development would be permitted. Under a system of administrative discretion, it is difficult for a national policy maker to be sure what values are being advanced at the local level and to take steps to establish consistency. An Interior Department task force later took note of this consideration in commenting that local "use tradeoffs can easily reflect mainly local values, or values of local BLM personnel. These values can also differ substantially from the priorities of the country as a whole. In addition, they are not likely to be consistent from one area to another. Under existing BLM planning system procedures, it is very difficult to know how land-use tradeoffs are made, because the values employed in decision making are almost always left implicit."[7] Another major concern was the inability of field managers to take account of cumulative impacts across many local decisions. It might seem quite reasonable to a local BLM manager to exclude coal leasing from a certain type of land which was important to his district. But, if applied across all BLM lands nationally, this kind of exclusion resulted in 30 percent of all federal coal land being removed from consideration for leasing, it might be another matter altogether.

The BLM recognized these disadvantages but considered them "a price worth paying in order to gain the advantages of wide local discretion to deal with widely varying individual circumstances at the local level."[8] But in 1977 the Carter administration decided to go ahead with development of uniform national unsuitability criteria. As will be recalled from chapter 10, Interior Secretary Andrus had made a key decision in October 1977 that the federal coal-leasing system should first determine, prior to receiving any industry nominations of tracts, the areas where coal development would be permitted. In November 1977 a task force was assembled with a charge to set policies for implementing this decision, specifically to develop "criteria for designating Federal land as suitable or unsuitable for coal leasing."[9]

The task force final report specified that its "goal was to develop a single set of criteria which would act as a screen to sift out Federal lands which should not be considered for Federal coal leasing. In order to achieve the goal, the Task Force recognized that the criteria must be specific so that the decision as to whether or not lands should be initially considered for leasing may be as unambiguous as possible. In effect, the criteria would serve as minimums for protection of certain land resources and values." By developing such criteria, they could be "applied in a fairly mechanical fashion to determine whether coal mining is suitable on any given piece of land. They do not require estimates of the value of the coal or making of comparisons of such estimates with other use values."[10] In other words, there would be no administrative discretion; the zones from which coal development would be excluded would be determined simply by the physical characteristics of the land and surrounding environment matched against the given criteria.

Zoning Out Coal Development

The unsuitability task force ultimately proposed twenty-two specific unsuitability criteria. For example, one criterion proposed that "Bald and Golden Eagle nests that are determined to be active and a buffer zone of land included in a ½ mile radius from the nest are areas which shall be excluded from coal leasing."[11] Other criteria provided similar protection for scenic and historic areas, raptor nesting sites, migratory bird habitats, wetlands, floodplains, and municipal watersheds. Several of the criteria, such as exclusion of mining in alluvial valleys or in endangered species habitat, came directly from recent legislation.

BLM land managers would still be free to designate further areas unacceptable for leasing because of conflicts with other valuable uses not specifically protected by any unsuitability criterion. But these decisions would be up to the discretion of the local BLM manager, unlike the unsuitability criteria. As the Task Force noted: "This method for determining availability of lands for coal development differs substantially from applying a given set of pre-established rules. It involves case-by-case balancing of coal development values, either subjectively or empirically, with other use values in order to maximize overall economic values obtained from public lands."[12] Under this procedure, for example, a campground or other desirable recreation area might be eliminated from consideration for leasing. However, the general expectation was that the greater part of the acreage excluded from leasing would result from the formal criteria.

Public reaction to the proposed unsuitability criteria was of two types. There was considerable acceptance that such a zoning system could prove useful in coal leasing. Environmentalists had long pressed for uniform national standards of environmental protection, but industry also saw a zoning approach as offering helpful advance guidance. It might aid industry to avoid investing too much in areas where either federal leasing would eventually be precluded altogether or there would be a strong presumption against it.

However, not surprisingly, there was much less agreement about the specific criteria—i.e., just where coal development should be zoned out. Environmentalists saw the criteria as too vague and favored tighter standards which would remove wider areas from availability for leasing. Industry was fearful that the unsuitability criteria, combined with other reasons for excluding coal development, predisposed the land-use planning system against coal mining and would preclude the leasing of urgently needed federal coal:

The process itself, and the sequence of the decisions in the Program, systematically gives precedence to all other articulated environmental, social and natural resource development policies. Land management decisions would be required to be made in the absence of adequate information concerning the nature or desirability of federal coal resources. Indeed, the recognition of the relative importance of such resources in comparison with other competing environmental or social values is specifically precluded throughout this stage of the planning process.[13]

Industry generally suspected that the unsuitability criteria would introduce further major obstacles in a coal-leasing system already regarded as excessively complicated and bureaucratic, and involving too direct a hand of the federal government in basic coal-production decisions: "This system will not produce a smooth flow of factual inputs leading to a well reasoned decision. Instead, the system is rife with potential for delays, conflicting inputs and recommendations, and would be likely to frustrate the Department's stated intent to expedite land use decisions."[14]

The Reintroduction of Administrative Discretion

It will be recalled that the basic decision by Interior Secretary Andrus to designate specific areas as acceptable or unacceptable for coal leasing reflected the difficulty of making a tradeoff among environmental and coal-development values. If decision making were costless, any coal tract of interest would be closely examined both for its coal development value and environmental consequences. From the pool of all such tracts, the few most desirable would be selected. But the very high information and other expenses of such a system required that a two-stage procedure instead be introduced: First, exclude those areas which clearly have high environmental cost, and then in the next stage make further determinations of relative tract desirability with a heavy weight placed on coal development value. As long as the area removed from coal leasing in the first stage was not too large a part of the total federal coal domain (itself huge), this system offered the reasonable prospect of bringing forth high-quality coal tracts which would have few if any serious environmental problems—all at a reasonable decision-making cost.

However, this two-stage system had the rigidities inherent in zoning—in fact, any regulatory approach applied with broad categories. It was open to criticism from both industry and environmental groups because its inherent inflexibility would produce the wrong decisions in some cases. Especially from an industry standpoint, some tracts with high environmental costs, but even more exceptional coal value, might well be passed over. A full balancing of coal-development value and environmental cost might have ranked these tracts among the best overall tracts despite their environmental problems. On the other hand, certain tracts with moderate environmental problems might survive the unsuitability screen and be leased ahead of tracts with fewer environmental problems and which, on an overall balancing of coal and environmental values, were preferable. Environmentalists would object to this possibility.

The justification for tolerating such mistakes would be simply that the costs to society of finding the mistakes would exceed the benefits. (The benefit of finding a mistake is the value of the gain—or savings—resulting from improvement in the decision made.) Perfect decisions are clearly a utopian ideal ruled out in the real world by the costs of making decisions, as well as other reasons. Thus, taking account of all kinds of social benefits and costs, including information and administrative costs of decision making, a system which makes a certain number of mistakes will almost always be the best choice. But it is difficult for public agencies to make such an argument; the public does not look kindly on explanations that an agency has deliberately chosen a course of action which it knows in advance will produce errors.

Rather, the pressure is much greater to assert that the system will give the right answer all the time. The easiest way to do this is to maintain wide administrative discretion; for any potential mistake, the maintenance of discretion always allows for its correction. To a critic who says "This will turn out wrong," the answer can always be given, "No, if you are right, it will be corrected." Hence, because it offers nothing to shoot at, administrative discretion is a main line of defense against before-the-fact criticism; agency critics are relegated to a second-guessing role after the fact.

The strength of such pressures for greater administrative discretion was evident in the evolution of the Interior Department's unsuitability criteria. The original objective had been uniform national criteria specified in advance. But this objective was gradually eroded as it became apparent that such rigid criteria would sometimes produce the wrong decision. For example, as it studied the matter, the Interior unsuitability task force "soon recognized that as the specificity of the criteria increased, the potential for arbitrary impacts also increased."[15] The difficulty was illustrated by the case of the proposed unsuitability criterion providing a buffer zone around bald- and golden-eagle nesting sites:

> It is extremely difficult to specify *a priori* the appropriate size of the buffer zone. Site specific data on topography, location of proposed mining activities, etc., should also be considered by the land managers when deciding how best to protect the eagle nest site from mining disturbances. Nevertheless, in an attempt to assess the impact of buffer zone requirements and to strive for as much specificity as possible, buffer zones were initially incorporated into the criteria. The size of the buffer zones were arbitrarily selected. For example, a buffer of ¼ mile radius around "active" eagle nests was ultimately proposed for field testing. A 1 mile buffer zone was initially proposed. . . . In some cases, a smaller buffer zone that that specified by the criteria would provide

an acceptable degree of protection while impacting much less coal. Thus, any specifically defined buffer zone has the potential to impact more coal resources than may be necessary. This potential problem can occur whenever a precisely defined criterion is applied to site specific data.[16]

The task force thus sought a compromise: "The problem . . . was to provide appropriate flexibility in the criteria so that local discretion may be exercised within specific bounds. The task force attempted to achieve this through the use of exceptions, where appropriate, and by field testing the criteria so that the proper balance point between resource protection and coal availability was established whenever such flexibility existed in the enabling statutes or policies."[17] With respect to bald and golden eagle nests, for example, four exceptions were provided to the criterion that a buffer of ¼ mile would be unsuitable:

A lease may be issued if:

1. Mining can be conducted in such a way and during periods of time that eagles will not be disturbed during breeding season.
2. A permit or special approval is granted by the FWS to allow the eagle nest to be moved.
3. During the non-breeding season mining can be conducted within the buffer zones.
4. Buffer zones may be increased or decreased if it can be determined that the active eagle nest will not be adversely affected. Consideration of availability of habitat for prey species shall be included in consideration of the buffer zones.[18]

The practical effect of these exceptions was to reintroduce wide administrative discretion for the BLM field. Many of the other unsuitability criteria also involved a number of exceptions of similarly broad scope. As it was evolving, the original intent of uniform national standards was being eroded, and the old, supposedly displaced approach of wide administrative discretion in the field was being revived. Further contributing to this effect was the vagueness and ambiguity of some of the unsuitability criteria themselves. For example, one proposed criterion stated that "fish and wildlife habitat for resident species of high interest to States that is essential for maintaining priority wildlife species as agreed by the State wildlife agency and Federal land management agency shall be excluded from leasing."[19]

A Further Retreat from Zoning

The greatest fear of the coal industry was that the discretionary system actually coming into being might be worse than no unsuitability criteria at all. It was unclear how the possibility of exceptions could be examined over the broad areas where coal mining was under consideration. To decide whether an exception applied seemed to require a site-specific examination. A company interested in mining at a particular site would have to put forward for discussion a proposal for limiting any adverse environmental impacts. If the proposed procedures seemed adequate, then mining would be allowed under the exception. The catch was that the Interior Department was committed to deciding on acceptability for mining before any specific industry proposals could be seen. Government officials, themselves, were not likely to make much effort to advance novel or complicated measures for preventing damaging environmental consequences.

Some of the budget officials of the Interior Department were equally concerned about the possible cost, as exceptions proliferated and site-specific examinations might be necessary to make a final unsuitability determination. If site-specific examinations had to be made over wide areas, the cost might become prohibitive. The question arose as to the actual wisdom of attempting to determine unsuitability in advance of specific proposals for mining. Costly checks might be made where later there might not even be any interest in coal mining in an area. The original rationale for unsuitability criteria had been to establish a pragmatic two-stage procedure for minimizing information and administrative costs, which now seemed to be slipping away.

Department budget concerns, added to the weight of outside criticism, led to the creation of a new task force to examine the data requirements for unsuitability determinations and possible ways of organizing the process to make these determinations more efficiently. The two major objectives of the task force were to "establish principles and guidelines which would assure that sufficient data are collected at the right time and place to support sound decisionmaking and would avoid, to the greatest extent practicable, collecting additional data of an intensive nature on areas which will not be leased for coal development" and "to recommend measures for achieving effective coordination of data planning, collection and sharing among various Bureaus involved in the coal management program, and thereby avoid costly duplication of effort."[20]

The coal-data task force now recommended a "staging strategy" which would be "more cost effective than non-staging strategies because it provides for the sequencing of surveys so that expensive surveys are undertaken only when necessary (only when such surveys are needed to increase the confidence of the findings, and only on the reduced acreage at later stages)." At each stage a crude benefit-cost test would be applied to determine how much data and other information would be collected. The benefit would be the improvement in the accuracy of the decision. In this case, the decision was whether a particular unsuitability criterion was satisfied in a certain area. The task force emphasized that, despite much popular belief to the contrary, a standard of absolute certainty should not be sought: "It should be noted that it will be almost impossible to have 100% certainty of an area's acceptability. Even if it were possible, making 100% certainty a decision rule would not be good public policy, because it would probably result in the non-development of rich coal areas with no unsuitability problems, just because of the time and expense of achieving 100% certainty."[21]

Applying the economics of data collection, at each stage more intensive data collection would be focused on the ever narrower areas still under consideration for coal leasing. The task force characterized the following overall data collection effort that would result:

> The incremental data collection (especially with respect to unsuitability criteria) at the land use planning stage can be characterized as being designed to cover large areas of land (e.g., one to two million acres of land annually) at relatively low cost on a per acre basis through use and analysis of color infrared aerial photography, low level aerial reconnaissance and limited ground surveys. It is expected that as a result of this first stage of incremental data collection and screening at land use planning, areas will be identified as either being acceptable for further consideration, or as being unsuitable because of the applicability of one or more unsuitability criteria. Of those areas identified as acceptable for further consideration, a portion will need further data collection beyond the first stage in order to increase the confidence level of acceptability.
>
> Those areas needing further data and screening will be identified and will be subject to further data collection of a more intensive nature at the activity planning stage, if they contain delineated tracts. This second stage of data collection and screening can be characterized as being designed to examine smaller areas (e.g., 200,000 acres) at higher costs on a per acre basis. For example, much of the expenses at this stage will entail on the ground investigations or drilling operations.[22]

Thus, there would now be a preliminary screen applied to rule out particularly obvious areas violating the unsuitability standards and where there was no real possibility of finding an exception. But most areas would probably have to be left for "further consideration." Such consideration might well occur after specific tracts had been nominated by industry and be limited to the area of the tract. In short, rather than creating zones well-defined in advance, the emphasis was shifting rapidly back towards case-by-case examination of individual tract circumstances, i.e., a system of wide administrative discretion.

Hence, in the end the sharp departure of the Interior Department from the EMARS II procedures for making unsuitability determinations proved to be short lived. Cost and other constraining pressures drove the Department to return to the original tract-by-tract emphasis. Outside the Department, even environmental groups were coming to think that this approach might be logical. Indeed, a few even suggested that the Interior Department would do well to return to the previously spurned EMARS II proposal to start off with industry nominations. In this way data gathering and other studies would efficiently be focused on the specific tracts of immediate concern:

The environmentalists admit their support for an early expression of industry interest in leasing represents a significant departure from the position they have taken in the past. Nevertheless, several reasons are presented for their new position:

"1. . . . One of the principal problems in the entire planning process [is] . . . the lack of generally available information for most areas that are potentially suitable for mining. As a result, the unsuitability criteria are allegedly not applied very effectively and early expressions of industry interest, 'would focus efforts to gather the information necessary to effective land-use planning and application of the criteria.'"

"2. . . . Although the industry is presently free to identify areas of interest before the activity planning stage and while still in the land-use planning stage, a formal request for industry's expression of interest defined 'according to townships' will insure that the unsuitability criteria are applied to an area large enough (larger than the area of specific industry interest) to allow an evenhanded application of the criteria, 'but still smaller than the total areas of medium and high coal development potential as identified by the USGS in a coal region.' Supposedly, then, data collection efforts can be more focused."[23]

The coal-data task force represented a new type of economic study that is likely to grow in importance. The subject was the economics of government decision making. As government has spent more and more for environmental impact statements, land-use plans, regulatory reviews, and other analyses, the total cost of such analyses has skyrocketed. In a few cases the analysis of a project has even cost as much as the project itself. Expenses for information gathering, proposal formulation, public review, and other procedures in decision making should be regarded as part of the overall investment in the public lands. Like other investments, they need to be subject to an analysis to be sure that the costs are not greater than the returns. Recognition of this need has been hindered by traditions of achieving full comprehensiveness in planning or complete ratonality in decision making which, ironically, leave out of the picture a key planning or economic factor—the cost of decision making itself.

Federal unsuitability designations were not the only mechanism for excluding an area from coal development. Congress had also given this authority to private surface owners in providing that they must grant their consent before underlying federal coal could be mined. In this case as well, the primary determination would occur tract by tract, according to the discretionary decisions of individual surface owners. In areas of privately owned surface, the requirement to obtain consent effectively made it impossible to plan zones in which permission for coal development could be assured by the Interior Department. It was another example of how government policy objectives could conflict, with one legislative mandate frustrating the achievement of the zoning system which Congress had also mandated—if with less than full awareness.

By the 1970s more and more professional land-use planners were rejecting the map making and formal blueprint traditions of their past training.[24] Instead, planning was coming to be seen much more as an exercise in incremental analysis of policy issues as they emerged. It was futile to try to make too many decisions in advance. Circumstances changed too fast; policy makers would want to make today's decisons with today's facts. Indeed, policy makers themselves came and went. In making decisions well ahead of time, it would be impossible to know exactly what the true circumstances would be later on; hence, there would always be some mismatch between the focus of earlier studies and currently emerging issues. Many analyses done too far ahead of time might be wasted if the issues for which they were intended never materialized.[25]

Another key change was an explicit acknowledgment that planning was a part of politics. It was not practical to try to bind future political leadership to the planning decisions of current officials. Moreover, current plans could not be made by experts working apart from political influence. But given the heavy demands on the time of top political leaders, it would not be practical to ask them to make too many decisions. Indeed, politicians would very often refuse to make decisions well in advance for fear of needlessly provoking political controversies.

The evolution of land-use planning and the development of the program for federal coal leasing generally provided a good case example in support of the new planning wisdom. But the coal program had to learn its planning lessons through trial and error. Belief in the virtues of formal planning was still very strong in public land management. The radical change in the role for planning proposed in the 1970s threatened the foundations of existing public land laws and institutions. Derived from conservationism, the essential justification for public land ownership had long been public management in the public interest for the good of all the people, as shown through scientifically rational planning. If most of these themes were really myths, what would be the future of the public lands?

V. Two Interpretations of Welfare-State Liberalism

14. Interest-Group Liberalism in Practice

If conservationism was failing its test, the question was what to do. What if scientific rationality and expert analysis could make only a limited contribution to many, maybe most, government decisions? What if good intuition and speculative hunches played a major role? What if most government decisions normally involved value judgments and other subjective considerations? What if government decisions would have to be made through the unruly procedures of democratic politics? What were the alternatives?

The development of the federal coal-leasing program was hardly the first occasion for the examination of such questions. Indeed, the fact that the public lands remained mostly outside the national spotlight from World War I to the late 1960s tended to insulate them from the impact of a wide rejection of progressive ideas among intellectuals after World War II. Another key factor was the unusual depth of progressive roots in public land management. As discussed earlier, the original retention of the public lands in public ownership was largely a product of the conservationist offshoot of progressivism. The institutions of public land management were formed largely under conservationist influence. The successes of conservationism were among the chief glories of the entire progressive cause. Probably more than any other current responsibility of the federal government, public land management has its origins in progressivism.

By the 1950s, however, a different ideology, sometimes called "interest-group liberalism," had largely supplanted progressivism as the public philosophy. This newer ideology elevated the practical politics of interest-group accommodation, as followed by President Franklin Roosevelt, into a formal political system. As conservationism showed signs of not working well in practice, the Interior Department increasingly looked to the ideas of interest-group liberalism to provide a sense of purpose and direction for its coal management.

Interest-Group Liberalism

By the end of World War II an active revolt had sprung up against the main tenets of progressive ideology. In 1946 Herbert Simon (who was to win a Nobel prize in economics in 1979), characterized the existing principles for scientific public administration as actually little more than a few "proverbs"—and mutually inconsistent at that.[1] But the most insightful and damaging critique was that of Dwight Waldo in his classic study, *The Administrative State*: "It must be reported that, with few exceptions, the notions of science and scientific method held by the writers are unable to withstand critical examination." Furthermore, Waldo found that much that is allegedly scientific in public administration was actually little more than common sense. But "'scientific method' is not, as often thought, identical with an extension of 'common sense.'" He similarly rejected the idea that government could be divided into separate political and administrative activities: "Either as a description of the facts or a scheme of reform, any simple division of government into politics-and-administration is inadequate. As a description of fact it is inadequate because the governing process is a 'seamless web of discretion and action.'"[2]

After the trust-busting efforts of the progressive movement, a more favorable attitude towards large business organization had been developing for some time. Indeed, the National Recovery Administration for the New Deal had been virtually a government-sponsored attempt to form large industry cartels, although it was quickly struck down by the Supreme Court. The new world of the large private corporation received the blessing of many U.S. intellectuals in the 1950s and 1960s. The writings of John Kenneth Galbraith, in particular, showed an acceptance of the inevitability of large organizations and the major benefits they offered in planning and coordination.[3] Contrary to progressive expectations, the leading models of planning in the 1940s and 1950s were not provided by government, but by large private corporations such as General Motors or IBM. The Weyerhaeuser Corporation was proving much more skillful than the Forest Service in achieving precisely the kind of scientific management of timber resources which Gifford Pinchot had advocated.

By the 1950s there had also been much greater actual experience with government administration, as a result of the New Deal and the general growth of the welfare state. This experience provoked much greater skepticism about any expectations of great skill in government management. The "scientific administration" of the old progressive administrative theory too often looked in practice like "inept or inert bureaucracy." A leading welfare-state proponent, the economist Gunnar Myrdal (a Nobel prize winner himself in 1974) was very concerned that "bureaucracy, petty administrative regulations, and generally a meddlesome state should not be the signum of our vision of a more accomplished democratic Welfare State."[4] He urged a minimizing of direct government commands and, wherever possible, resolution of resource conflicts in the private sector, although mostly by political bargaining rather than market competition.

In 1951 David Truman published *The Governmental Process*, a highly influential reformulation of how American politics actually worked. The Truman thesis rejected the basic progressive conviction that government could identify and then act in a well-planned fashion to achieve any single "public interest." For Truman such progressive beliefs were simply myths that bore little relationship to the real world of government; they were important, but only as part of the "data of politics." If large enough numbers of people really believed that there was a public interest, this fact in itself was significant, even if lacking in any objective validity.

> Many . . . assume explicitly or implicitly that there is an interest of the nation as a whole, universally and invariably held and standing apart from and

> superior to those of the various groups included within it. This assumption is close to the popular dogmas of democratic government based on the familiar notion that if only people are free and have access to "the facts," they will all want the same thing in any political situations. . . . Such an assertion flies in the face of all that we know of the behavior of men in a complex society. Were it in fact true, not only the interest group but even the political party should properly be viewed as an abnormality. The differing experiences and perceptions of men not only encourage individuality but also . . . inevitably result in differing attitudes and conflicting group affiliations.
>
> Assertion of an inclusive "national" or "public interest" is an effective device in many . . . situations. . . . In themselves, these claims are part of the data of politics. However, they do not describe any actual or possible political situation within a complex modern nation.[5]

Instead, Truman found that the actual governmental process was characterized by competition among numerous groups as each sought to advance its own special interest. Radically departing from progressive views, the key role of the government official is to mediate among these groups and produce a workable interest-group compromise:

> Where compromise in the legislative stage is the alternative to temporary failure and where the imperative to compromise is accepted by some participants as a means of avoiding the open frustration of expectations widely held in the community, the terms of legislative settlement are almost bound to be ambiguous. Such compromises are in the nature of postponement. The administrator is called upon to resolve the difficulties that were too thorny for the legislature to solve, and he must do so in the face of the very forces that were acting in the legislature, though their relative strength may have changed.[6]

Other prominent students of American politics found a similar large discrepancy between the ordinary citizen's belief that the government could identify and pursue some single national interest and the way that the political-economic system was actually working. At a Brookings Institution panel in the early 1960s a well-known scholar found that "there is much in the democratic credo that cannot be taken literally. It is akin to trying to talk French by pronouncing the words the way they are spelled: such utterances would confound any Frenchman. Citizen participation, responsible government, the public interest, the democratic consensus, rule by public opinion—the operating significance of such sounds and symbols is acquired only by long habituation. We grow to learn which of the letters are silent."[7]

Unlike earlier progressives, Truman and other more recent commentators did not find much fault in a governing process in which the outcome was largely determined by special interest competition. Indeed, as long as it was reasonably balanced, the resulting equilibrium of interest-group pressures would be satisfactory. As Galbraith argued, as long as a circumstance of "countervailing power" among major social interests was maintained, the nation's needs would be served. It was a kind of invisible hand in the political domain; the pursuit by various groups each of its own special interests would produce a final result in the overall social interest.

As Theodore Lowi characterized it, such views consitituted the "new public philosophy, interest-group liberalism" which became the "foundation of the Consensus" for governing in the period after World War II. According to Lowi, interest-group liberalism followed directly from a few simple steps in reasoning:

> (1) Since groups are the rule in markets and elsewhere, imperfect competition is the rule of social relations. (2) The method of imperfect competition is not really competition at all but a variant of it called bargaining—where the number of participants is small, where the relationship is face-to-face, and/or where the bargainers have "market power," which means that they have some control over the terms of their agreements and can administer rather than merely respond to their environment. (3) Without class solidarity, bargaining becomes the single alternative to violence and coercion in industrial society. (4) By definition, if the system is stable and peaceful it proves the self-regulative character of pluralism. It is, therefore, the way the system works and the way it ought to work.[8]

Indeed, the "pragmatism" and disavowal of ideological commitment of many Americans on closer inspection often turned out to mask a specific ideology—that of interest-group liberalism. To be "sophisticated" meant to recognize that politics was not a matter of pursuit of the public interest; instead, it was a direct competition among interest groups for benefits in the political arena. People "in the know" understood that this was the way the system really worked—whatever the lessons of high-school civics. The reaction to this fact of life could range from deep cynicism to active approval. But interest-group liberals ultimately believed that in any case no better system was available.

Interest-group liberalism was actually an American version of a broader international movement. Reflecting Swedish experience, Myrdal considered that in the fully realized welfare state, resource allocation would be accomplished through a process of bargaining among the leading social interests. Wages might well be set through nationwide bargaining between business and labor. Myrdal expected that "such forms of general income settlements among the main organized interest groups in a national community will gradually become the rule. All prices and wages and, in fact, all demand and supply curves, are then in a sense 'political.' We are as far away as possible from the 'free market' of liberal economic theory. The government and the administration . . . will then gradually find it as important to lead the negotiations and to control the compromises between the nationwide organized power groups, as it is to lead parliament itself."[9]

Reflecting the more homogeneous nature of Swedish society, Myrdal put less emphasis on the competitive aspect of interest-group bargaining than the American interpreters. He is optimistic that bargainers will be "rational" and recognize a strong common interest to pull together to achieve widely shared social goals. Thus, his prescription for inflation is that it "can only be solved in a fully satisfactory way by raising the general level of education, intensifying still more the act of participation of the people in decisions on all levels, strengthening very much the awareness which all should have of the common interest that the price level should not get out of hand, and creating thereby the basis of understanding and solidarity required for the national planning and coordination of all markets."[10]

According to the tenets of interest-group liberalism, the government does not generally seek to advance any particular substantive policies. The various interests will themselves espouse their own often conflicting policies. The key government skills involve brokering an acceptable bargain to resolve the disagreement. Carried to its ultimate logic, ideas become nothing more than tools of interest-group manipulation; a convincing ideology is merely a clear device for creating a better bargaining position—a "line" to win greater political favors. The success of government lies in its ability to head off and diffuse social conflict, to foster an attitude of cooperative search for mutual accommodation, and to reach settlements that do not come unglued. This was very far removed, indeed, from the ideals and principles of progressivism.

Interest-group liberalism brought a whole new attitude toward special interests; instead of denying their legitimacy, and seeking to banish them, interest groups were welcomed as central elements of the governing scheme. The public interest thus came to be defined by mechanical procedures; if they were fair to all affected parties, whatever substantive result was achieved by definition would become the public interest. Procedural rationality replaced substantive rationality as the paramount government concern—a major virtue whenever substantive agreement is very hard to obtain.

A step back toward progressive ideas would allow for the government—or at least some segments of it—to function as the representatives of particular social interests. The government thus might serve as the spokesman for the poor, the disabled, or any other group likely to be disadvantaged in political competition. It similarly might represent diffuse national taxpayers or all those broadly concerned with efficient use of social resources—who otherwise would find it very difficult to organize to express their concerns in the political process. This version of interest-group liberalism, however, raises the question of how the government can legitimately write the rules of the game, then play in the game, and ultimately be the final judge.

A federal coal program following the prescriptions of interest-group liberalism would aim to create appropriate settings in which all concerned interests would come together to negotiate satisfactory leasing levels and other policies for federal and western coal development. Presumably, any group with a reasonable interest in such matters would have a right to some voice in federal coal decisions. The concerned interests, at a minimum, would include coal companies, western coal states, environmentalists, local governments in western coal areas, and coal-burning utilities. The decisions concerning the location, rate, and other aspects of federal coal production would be worked out among these interests. The federal government would function to some extent as an even-handed mediator and also, conceivably, as itself an advocate for the diffuse interests of national energy users who might find it difficult to be represented in any other way. Among other activities, government would provide facts and figures, technical assistance, and other useful background information to assist in reaching an interest-group settlement to greatest overall satisfaction. The policies for federal coal development would, in essence, be politically determined.

Learning the Ways of Washington

The verdict of history on the Carter administration may be that it took the American myths too literally. Other administrations had taken care not to contradict myths such as "the public interest," or had even actively manipulated them for partisan gain, recognizing their power as symbols. In David Truman's terms, these administrations considered them important "data" for politics. However, they had recognized that there was often a wide gap between the way the political process was supposed to work in the popular mind and the way it really worked.

The Carter administration came into office determined to banish traditional political deals and to turn matters over to the experts. It would be a revival of progressive ideals. President Carter, perhaps because he had a scientific background himself, reflected habits of thought and a political approach directly in the line of Presidents Theodore Roosevelt and Woodrow Wilson. There was a correct answer—at least implicitly the scientific answer—to most policy questions. It was only necessary to assemble the leading authorities and have them spell out this answer. The Carter administration thus turned out a series of comprehensive policy prescriptions prepared by expert task forces in major national problem areas, beginning with the national energy plan.

On a lesser scale, the approach to federal coal management followed this pattern. The politically tainted efforts of previous administrations to establish a federal coal program would be discarded at the outset. Instead, a brand new comprehensive program would be developed—this time truly in the public interest—and particularly excluding the special influence of the coal industry. Task forces of Interior Department personnel knowledgeable in coal matters were assembled to formulate the new federal coal program from the ground up. Indeed, the whole effort would ultimately require two and one-half years, not being completed until June 1979.

In the case of the federal coal program, unlike some other policy areas, the Carter administration did not have to go to Congress for any new legislation. Although seldom previously put into practice, the existing legislation had origins in progressive and conservationist ideology; it thus formally provided for land-use planning and other mechanisms of technical decision making by experts. The previous chapters have described the frustrations as the Carter administration initially sought to manage federal coal in an expert, rationally planned fashion. Gradually, however, the Carter administration abandoned this approach. By 1980 interest-group liberalism had emerged as the basic philosophy of the federal coal-leasing program.

To be sure, important elements of interest-group liberalism were found in the Carter coal program from the very beginning. Government frequently mixes elements from different ideologies, partly because officials do not all share the same outlooks. Moreover, like most people, individual officials themselves may have ideological conflicts or incompatibilities within their own positions or arguments. More generally, in recent years American government has often put forth a progressive public face, while actually acting in an interest-group liberal manner.

A Broker among Federal Coal Interests

When the Carter administration came into office, its spokesmen typically referred to the "public interest" in a way that suggested the traditional conservationist understanding. The President's Environmental Message in 1977 directed that federal coal leasing should be undertaken "in a manner that fully protects the public interest."[11] Guy Martin, the newly appointed Interior Assistant Secretary for Land and Water Resources, asserted that the situation he inherited was "intolerably contrary to the national interest."[12] Yet, on closer examination, the concept of the public or national interest offered by the new Carter administration was often really derived from interest-group liberalism.

On some occasions the influence of interest-group liberalism was plain to see. Thus, new Carter administration officials considered that the failure of federal leasing under EMARS II was not so much in the substantive results achieved. Indeed, almost no federal coal had been leased by the Interior Department since 1971—from the Carter administration viewpoint not so bad in light of the huge reserves in existing leases and the previous public opposition of many of its incoming appointees to leasing. Rather, the greatest failure was that by appearing to be biased towards the coal industry and showing little heed for the interests of western states, environmentalists, local governments, ranchers, and other groups affected by federal leasing, the Interior Department had created a storm of opposition to its own policies. It had greatly aggravated interest-group conflicts rather than diffused them, encouraged recalcitrance instead of cooperation, and in short, had thoroughly mismanaged its proper role of brokering an agreement on a coal program acceptable to all the major interest groups affected.

As Assistant Secretary Martin remarked, "I can only say that the previous system was producing a lot more conflict than it was coal."[13]

Martin considered that there were a number of legitimate interests: "State governments have provided thorough and competent criticism of the old system. The coal industry, both users and producers, have made clear the need for certain, understandable, predictable standards. Agricultural, environmental, and Indian interests have been specific in their comments on Federal coal policy." But the problem was that "the one element common to most of these interests is unhappiness with the present system. . . . My personal feeling is that we started about as close as possible to the bottom." A main part of the solution was to get everyone back together to achieve a satisfactory compromise. The new leasing system would treat each of the interests as "respected . . . partners in energy development." Although previously many interest groups had been considered "obstacles," Martin now stated that in the future they would all be regarded "as real and legitimate interests." Once there was a genuine determination on the part of the Interior Department to reconcile differing interests, progress toward coal development would occur much more rapidly: "Most important, I believe, is this Administration's belief that it is possible to reconcile the conflicts which had stalemated past efforts to produce coal."[14]

Gradually, these themes became predominant in Interior public statements concerning the coal-leasing program. In announcing a new program in June 1979, Interior Secretary Andrus now indicated that the ability to smoothly resolve interest-group conflicts was perhaps the most important program objective of all: "The goal of the program is to allow for progressive development of our vast Federal coal reserves in ways acceptable to the many competing interests." He emphasized that the coal program would provide "full participation . . . for State and local governments, the minerals industry, environmental groups, ranchers and the public to help determine policies on coal development, land use and land reclamation."[15] In October 1980, Secretary Andrus announced a federal coal lease sale for January 1981—the first in ten years. He considered that the strong Interior commitment to effective brokering among the various affected interests was the prime reason for success in finally getting coal leasing resumed: "To put this vital energy program in place, right on the original schedule announced by the Department is a major achievement. To succeed, we had to overcome a history of litigation, a history of poor Federal coal management, and an inherent conflict with the Western States, where the significant Federal coal deposits are located. We also had to reconcile the many concerns of public interest groups. The new leasing program has won the genuine support of the states and the general public by using a balanced approach that recognizes the legitimacy of all uses of the Federal public lands."[16]

Other top Interior officials defined the success of the new coal program in similar terms. As former NRDC staff member and top Interior official commented, "The Republican Administration had halted Federal coal leasing in the early 1970s, but its attempts to devise a new leasing program had been stymied by opposition from Congress, the Courts, ranchers and many others in the West." But matters now were greatly improved because the Interior Department had designed "a new coal-leasing program which has won praise from nearly all affected interests."[17] Reflecting the usual practice of disaffected interests to resort to court challenge, another official commented, "We've been almost litigation free, I think it's a hell of a success story."[18] Representatives of state governments confirmed that from their point of view much greater co-operation had indeed been achieved. The coordination mechanisms between the federal government and the states that had been established "serve an extremely worthwhile purpose and are operating with a minimum of difficulty to date."[19]

To be sure, the coal industry was not as happy. In fact, the question can be raised whether Interior had actually resolved interest-group conflicts or merely shifted sides, changing the disaffected party. In 1980 one neutral observer of the program commented: "Interior has a . . . practical concern: how to avoid lawsuits by states and environmentalists opposed to the program. This probably motivates Interior officials more than anything else, and makes them more eager to satisfy environmentalists and state officials than to guarantee plenty of land for the coal industry."[20] But if one interest group had to be unhappy, perhaps the coal industry was the wiser selection. Industry would always mine federal coal if it were made available; state governments and environmentalists had shown ample capacity and inclination to shut off the availability of new federal coal altogether—a genuine major threat.

Procedures for Negotiating Leasing Levels

As the design of the successor program to EMARS II got underway in 1977, Secretary Andrus stated that "the new Department of Energy . . . will make it possible for the federal government to provide much more precise estimates of the amount of coal we will need in future years. The Interior Department will be able to respond to these long term production goals."[21] This was in essence, a traditional conservationist theme. But by June 1978 matters were changing. In resolving the issue of how to determine leasing levels, Secretary Andrus rejected an alternative that seemed to reflect closely his earlier thinking, "Lease to meet DOE national production projections." He instead selected another alternative, "Lease to meet DOE production projections, modified by industry, state and local and other inputs."[22]

In this insertion of a brief phrase, "modified by industry, state and local and other inputs," the Interior Department took a major step away from conservationism and towards interest-group liberalism. Leasing levels would not be determined exclusively by DOE and Interior energy experts. Rather, the DOE position would be incorporated into the leasing system as the view of another interest—if perhaps a very important one. Interior would perform a balancing function among all the interest groups, including DOE. As the Department later explained, the new coal program would provide a "process which merges DOE regional production goals with advice from state and local governments, DOJ [Department of Justice], the coal industry, and other interests to determine leasing targets."[23]

By 1979 the Interior Department had become still more skeptical of the existence of an objective justification for the production goals DOE was generating. These goals were increasingly seen as the easily manipulable vehicle for DOE to put political pressure on the Interior Department to lease more federal coal. Neither DOE or anyone else really knew exactly how much western coal would be produced or how much federal coal was needed. But if DOE somehow determined that more leasing was generally desirable, it could simply change an assumption or find some other way to raise the production goal. In this regard, DOE might simply function as an intermediary for industrial groups who were pressuring it to put more heat on the Interior Department.

The final coal program regulations, published in July 1979, contained a key section on "Regional production goals and leasing targets."[24] This section prescribed an elaborate set of procedures for setting coal-leasing targets. The procedures were designed to ensure that every affected interest was amply consulted and would have no reason to complain that it had been denied a fair opportunity to influence the amount of federal coal eventually leased.

The first step under the regulations would be for DOE to state "proposed regional production goals." Within sixty days the Interior Department would comment, including consideration of "the national need for coal resources balanced against the environmental consequences of developing these resources." Following that, within thirty days the Secretary of Energy would issue his final goals, commencing the process of setting the leasing targets for federal coal. The DOE goals would now be reviewed in light of "state government, Bureau of Land Management state office, Indian tribe and regional development policies." Public hearings opened to wider groups would also be held. Considering the comments of such affected interests, the Secretary of the Interior would next publish "preliminary regional leasing targets" and make "any adjustments" to the DOE goals that he might decide "are necessary." A Federal Register notice would then be published for review of "coal and utility industries, agricultural and community organizations, environmental groups, Indian tribes and other concerned parties." Further consultation would follow with the Secretary of Energy and state governors concerning the preliminary leasing targets, especially "the potential social and economic effect on the state and region." The regulations finally get to the last step in the process: "Based on the consultation with state governors, consideration of the Department of Energy's final regional production goals, as adopted, and the comments received on these goals and the preliminary regional leasing targets, and the comments received, the Secretary shall adopt final regional leasing targets."[25]

Although the basic objective is to mediate an accommodation among all concerned interest groups, traditional conservationism does sometimes survive in the format into which the interest-group bargaining is pushed. Much like DOE, interests favoring more rapid coal development may well have to pursue their objective by seeking assumptions, forecasting methods, and a general analytical framework which yields their preferred conclusion. Interests opposing federal coal leasing similarly seek to gain acceptance for assumptions and an analytical framework which yields slower federal coal development. The public debate thus formally adheres in some cases to the old concept of a public interest established scientifically; but in fact, the process is more a series of interest-group bargaining sessions over the key assumptions and computing methods that it is known will determine a particular final result.

This is not to say, however, that each interest group does not decide its own position in part on the basis of demand studies or other technical considerations. Moreover, technical expertise is also necessary because the ability to succeed in the bargaining sessions may depend critically on skill in articulating interest-group objectives in a technical language and framework.

One of the key innovations of the new coal-leasing program was its creation of "regional coal teams." Each such team would consist of three federal government representatives and two state government representatives. But at least two of the federal members would be based in the western states and often reflect western public opinion—certainly be in close touch with it. The regional coal teams would serve as a conduit to the Secretary of the Interior for the various western interests concerned with federal coal leasing. The Secretary might find that his task would essentially be to balance, on the one hand, DOE and coal industry pressures for national energy development and, on the other hand, western concerns as relayed by the regional coal teams.

The regional coal teams were also given a key administrative role to coordinate lease sale planning in each federal coal region. The Secretary of the Interior necessarily reserved the final decision to himself, but the coal teams would play a leading role in setting overall targets, structuring lease sale alternatives, and generally organizing analytical documentation. Representatives of state government would become integrally involved in the coal-leasing process. The creation of the coal teams was a key element in efforts to enlist the active participation and cooperation of western states in the leasing program. The states for the most part responded enthusiastically to the team concept:

> Although at first glance, the Teams appear to be little more than a staff coordinating mechanism to implement the new federal program, in actuality they are a rather unique experiment in federal/state cooperation. . . . This author wholeheartedly supports the Department of Interior's decision to make the Teams the central focus of all policy and implementation activities for the new program. For those who have participated on or worked with other federal advisory committees, the unique character of these particular federal coal advisory committees (the Teams) is clearly evident.

The Regional Coal Teams, and thus the Federal Coal Advisory Board, are comprised of voting members representing the BLM, the state directors of the BLM, and representatives of the governors of each of the affected states. These particular individuals will ultimately retain direct responsibility for making major policy changes and implementing these changes within the federal establishment, and the state representatives have an additional responsibility to implement policy in the affected states. So, although the Teams are termed coal "advisory committees," they are advisory with a twist; i.e., the particular committee members are at once advisors and the principal persons responsible for implementing their own advice, subject always, of course, to the ultimate approval of the Secretary of Interior and individual governors. Moreover, establishment of the Teams and the authorization for the governors of affected states to nominate a voting member of the Team was the result of a realization that state government is not merely another member of the general "public."[26]

Striking the First Bargain

Before the issue of the exact amount of federal coal leasing had to be faced, the first question was whether any new leasing would occur at all. Indeed, resolving this matter took up almost all of the 1970s; it was not settled until January 1981, when the first regular sale of federal coal leases since 1971 took place. The decade-long bargaining required to achieve this sale gave an example of interest-group liberalism in practice.

Major economic realignments generally create patterns of winning and losing interests. The biggest losers from western coal development were eastern coal mining, especially the United Mine Workers, and the people living in the western coal areas whose traditional way of life would be disturbed. The visitors to coal areas often are from the major urban areas of the West, and to a lesser extent from the country as a whole. These visitors, coming mainly for recreational purposes, would also be likely losers. The biggest winning interests would be the direct participants in western mining, both employers and employees; the many secondary beneficiaries of greater economic development of the West; and the nation's energy consumers, who would gain access to a cheaper and more reliable source of energy. Conflicts of interest among these groups had to be settled in some fashion before federal leasing could get back under way.

The market is the major force for economic change in the American system; wherever greater profits can be achieved, market forces automatically push economic activity in this direction. But politically negotiated decisions show an opposite tendency—more often holding back change. The winning interests from change are frequently spread thinly and are amorphous, whereas the losing interests are more typically concentrated and easily identifiable. This circumstance tends to give the latter a greater political influence. In order to avoid a permanent stalemate, the resolution of interest-group conflicts sometimes takes a special market form: the winning interests agree to pay off the losing interests. Often the winners can easily afford to purchase consent of the losers, since they gain much more than the losers stand to lose. Because there is no good mechanism for direct cash transfers, and in any case they are likely to be ethically frowned upon (as bribery, fraud, etc.), payoffs to losers typically take less direct and more socially acceptable forms. Congressional logrolling, for example, is one of the most important institutions for this purpose.

In the case of western coal development the biggest payoff has been a large increase in coal revenues going to western states. The royalty rate on federal coal was raised by the Federal Coal Leasing Amendments Act of 1976 from less than 5 percent prior to 1971 to a minimum of 12.5 percent for surface mined coal. Moreover, the Act increased the state share of federal bonuses, rents, and royalties from 37.5 percent to 50 percent. State governments have also acted independently to capture major new revenues from coal development; Montana imposed a 30 percent severance tax in 1975, and by 1980 the Wyoming severance tax had climbed to 17 percent. Based on recent coal production forecasts, coal revenues to Wyoming in 1990 will probably exceed $250 million, and to Montana, $200 million. By comparison, in 1978 the total revenues collected by the State of Wyoming were just $671 million and by the State of Montana just $923 million.

Western states have claimed that the coal revenues received are necessary to pay for public infrastructure and other burdens imposed by coal development. But the evidence indicates that in a number of cases the revenues received will be far greater than required for this purpose. In a study of fiscal impacts of coal development in the Montana part of the Powder River region, three economists at Resources for the Future reported that "our own results suggest that a relatively small fraction of the severance tax receipts would meet the fiscal (as distinct from psychic) burden of local impact mitigation." For the state government "the fiscal impact . . . would be overwhelmingly favorable"; for local counties in the Powder River region it would be "very favorable."[27] Like Alaska with its citizen distribution scheme to rapidly dispense accumulating oil revenues, Montana has also been hard pressed to decide where to put all its new coal-tax money. Much of the revenue is presently going into a trust fund for future use.

A second way to placate potential losing interests from economic and social change is to provide them with greater insulation from such change. Society has many institutions for this purpose; guaranteed job tenure is a good example, protecting against cyclical employment instabilities, shifts in labor force composition, as well as declines in the skills and marketability of individual workers. As a group coal miners in the East sought protection against declining mining employment due to westward shifts in patterns of coal production. Congress moved to provide such protection in the Clean Air Act Amendments of 1977, significantly reducing the environmental advantage of western low-sulfur coal. Compared with a direct payoff, insulation from social change in this manner is often much less efficient, imposing a dead-weight loss on the whole economic system. In this case the loss resulted from the uniform national requirement to employ very costly scrubbing techniques even for clean coal that could already meet existing air quality standards without any scrubbing. Politically, however, it is often easier to impose a less conspicuous loss of this kind than it would be to make direct cash payments to losers (e.g., perhaps provide special bonuses or special unemployment benefits to eastern miners).

Another influential set of potential losers from western coal development would be existing ranchers who owned the private surface but not the underlying federal coal rights. Under law, prior to 1977, they had to be compensated for any damages due to development of federal coal, but the extent of such compensation and whether it would really fully make up for imposed disruptions was doubtful. A provision requiring formal consent of the private rancher was inserted into the Surface Mining Control and Reclamation Act of 1977 to take care of these potential losers. In this instance they could choose either to take a large direct financial payoff in exchange for their granting consent to coal mining, or instead could choose to be insulated from mining by refusing to give their consent. Recent reports indicate payments to ranchers of as much as $5,000 per acre or a 3 percent royalty on production of the underlying coal. Many individual ranchers are likely to receive well above $1 million for granting permission to mine "federal" coal.

The most numerous, if least concentrated, of those losing interests from western coal development were environmentalists and recreationists. Coal development might scar large areas of western landscape, contaminate and dirty rivers and streams, threaten wildlife populations, hamper visibility in national parks and other national scenic areas, and have other undesirable environmental and recreational consequences. Responding to such concerns, tight controls and a requirement for the reclamation of surface mined lands were provided under the Surface Mining Control and Reclamation Act of 1977, which also included strict new protections against mining impacts on water quality. The Clean Air Act Amendments of 1977 limited allowable deterioration in air quality in national parks and other nationally important areas. The Federal Coal Leasing Amendments Act of 1976 and the Federal Land Policy and Management Act of 1976 provided new, if less direct or secure, protection for wildlife and other uses through requirements for comprehensive land-use planning. The Interior Department's unsuitability criteria in 1979 sought to make such protection more explicit.

The coal industry and federal energy agencies were critical of the various measures taken to accommodate the interests of recreationists and environmentalists. These measures were said to be too expensive and to remove too much land from possible coal mining. For its part, the Interior Department considered such criticisms myopic. The coal industry was refusing to play the game straight; its idea of a bargaining process was everything for itself and nothing for opposing interests. If the coal industry were allowed to have its own way, it could well be its own ruin. As Interior Secretary Andrus explained in a 1979 memorandum to President Carter, "Even if DOE coal-related planning and environmental standards were substantially relaxed, no significant increase in production or use of coal would result. There would be some reduction in costs and some changes in location in coal production and use, but conflict and resulting delays for individual development proposals would likely increase. State-Federal relations would return to pre-1977 conflicts, and public acceptance of increased coal use would diminish."[28]

As western coal mining actually began to reach significant levels in the late 1970s, it started to create a new concentrated constituency of its own, thereby increasing its political clout. Western mining companies, workers in new western mines, suppliers of surface-mining equipment, purveyors of legal and financial services to mining, even environmental scientists employed to control damaging impacts—all had an interest in further western coal development. Combined with a more general desire for economic growth, reflecting harder times and the recognition by businessmen and workers in many fields that they probably stood to benefit, the political pressure for western coal development had heightened considerably by 1980. Finally, the OPEC price hikes of 1979 and the resulting renewed national concern over excessive dependency on foreign oil created strong additional pressures for more rapid development.

The coal-leasing history of the 1970s can thus be seen as a drawn-out negotiation over whether federal coal leasing should lend any support to western coal development. The interests with the power to hold up western coal development had to be dealt with, either by paying them off or meeting their concerns in other ways. It turned out to require ten years to reach a settlement acceptable to them, the coal industry, and the rest of the country (and the matter may still not be finally settled). The lawsuits, data gathering, EISs, planning, and long debates and controversy about the ideal federal leasing program can be seen as simply incidental byproducts of the lengthy negotiations taking place among interest groups.

An obvious question is whether such a prolonged negotiating process is too slow and too costly for resolving major social issues. One view is that the system in its curious ways worked; ten years is not too long to mull over a question as important as a major transformation in the social and economic system of the West. Many of the numerous studies and reports that formed a backdrop for the controversy were probably useful. There was enough federal coal already under lease that the costs of holding up further federal leasing may not have been all that large, and indeed, may well have been very small. Mines developed on existing federal leases could well have been as efficient or environmentally satisfactory as would the mines possible with new federal leases. There is no clear evidence that coal production was forced to shift from one coal region of the country to another because new federal coal was not available.

Nevertheless, the federal coal-leasing program cannot be considered in isolation. What if other programs which raised issues of similar national scope all required ten years to decide? Granted that western coal development is a big issue, there are still a large number of equally important matters before the nation, many of them controversial. Given the incremental nature of decision making, national policies must be formed in some logical sequence. If decisions naturally assume a sequential structure, so one must await the resolution of another, and if each decision by itself takes a long time, the overall time required for social decision making may become extraordinarily drawn out. The nation does not want to take decades to form its energy and other basic social policies.

The failure to lease federal coal in the 1970s probably did little to hold back the workings of market forces; western coal was effectively constrained by limited demand, not a lack of supply. Strong market forces in themselves tend to create powerful interests whose needs will generally be accommodated in some fashion. On the other hand, there is no special interest constituency for planning itself—only when the results are thought to be known beforehand and to favor an interest which will then push for more planning.

A planning system that is rational in its overall decision-making structure must limit the amount of time that can go into each individual decision. Overall national energy planning, as well as planning of coal companies, utilities, and other individual components of the national energy system, has very likely been unduly impeded by the protracted bargaining resulting in ten years of uncertainty over federal coal availability. Paradoxically, in light of the widespread demands for better planning that held up leasing, the greatest casualty of interest-group liberalism may have been the frustration of rational coal and energy planning.

Two obvious minimum requirements for a good plan are that it should not prescribe goals which are directly conflicting, nor should it recommend actions which are directly at cross purposes. Yet ordinary politics gives little importance to either of these requirements; indeed, the nature of interest-group bargaining virtually assures that they will be violated. A skillful mediator often tries to obscure conflicts among objectives; he seeks to achieve an acceptable balance among interest groups by eventually conceding a little bit to the goals and strategies of each group. Yet, it is in the nature of interest-group conflict that the various goals and proposed actions offered will very often be inconsistent; as a result, interest-group liberalism by its very nature will tend to produce actions and goals at cross purposes.

The difficulties of planning in an environment of constant political bargaining and compromise have attracted much wider commentary. The free-market economist, Friedrich Hayek (winner of a 1974 Nobel prize), considered that the chronic inability of democratic government to plan would eventually bring on demands for stronger central authority, leading ultimately to socialist dictatorship: "Agreement that planning is necessary, together with the inability of democratic assemblies to produce a plan, will evoke stronger and stronger demands that the government or some single individual should be given powers to act on their own responsibility." The problem is that "in the direction of economic activity the interests to be reconciled are so divergent that no true agreement is likely to be reached in a democratic assembly."[29] The American Theodore Lowi echoed a similar skepticism about the prospect of effective planning emerging from the negotiating processes of ordinary political give and take: "Liberal governments cannot plan. Planning requires the authoritative use of authority. Planning requires law, choice, priorities, moralities. Liberalism replaces planning with bargaining. Yet at bottom power is unacceptable without planning."[30]

Such observations point to one of the central dilemmas of modern society: how to reconcile democratic freedoms and politics with a desire for effective government planning. So far the two seem incompatible; the progressive thesis having proven unsatisfactory, the interest-group liberal antithesis doesn't look much better. One wonders whether there is any satisfactory synthesis.

Surface-Owner Consent

The difficulty of defining consistent goals and policies in the normal political process was well illustrated by the 1977 enactment of the surface-owner consent requirement of the Surface Mining Control and Reclamation Act. As noted above, under this requirement qualified private surface owners—essentially ranchers of more than three years duration—must give their agreement before federal coal beneath their property can be mined.

The previous year Congress had enacted the Federal Land Policy and Management Act, containing the dominant theme that the era of public land disposal had formally ended. Although land disposals had largely concluded as a practical matter in the 1930s, for the first time a clear statutory mandate was enacted putting the public domain lands permanently under federal ownership and management. The new law directed that the management of the public lands should be based on land-use planning. It affirmed the importance of receiving fair market value in the leasing or sale of federal resources. The basic intent of the legislation was to make possible the creation of a modern management system by which the broader public interests would be well served on the public lands.

Although the new consent requirement effectively conciliated surface-owner interests, it also placed major stumbling blocks in the way of implementing the objectives of the Federal Land Policy and Management Act. Prior to the surface-owner consent provision, the federal government as owner of the coal rights was entitled to mine its coal (or lease it to someone else) so long as it paid adequate compensation to the private surface owner for any loss in agricultural production or any other damages caused by surface mining. But, now, the previously unambiguous federal right to mine coal became a shared right; federal coal could not be mined without the joint approval of both the federal government and the surface owner. Since the main property right of coal ownership is the right to control coal development, the federal government had, in a real sense, taken on a partner; it had effectively disposed of a significant share of the property rights to its coal. Henceforth, the division of the revenues from developing federal coal would have to be determined by a bargaining or negotiating process between the federal government and surface owners. The surface-owner consent requirement was the most significant disposal of federal rights on the public lands since homesteading ended in the 1930s. It was enacted a year after Congress supposedly had put a final formal end to the era of disposal.

Because of its huge holdings and consequent much greater potential bargaining power relative to individual surface owners, the federal government may yet be capable of squeezing the bulk of coal revenues away from surface owners. But, after some heated debate, Congress voted down attempts to place any explicit limits on payments to surface owners for their granting of consent to mine.

The requirement for surface-owner consent also creates a major barrier to competitive bidding in federal coal-lease sales. If one coal company has already purchased the exclusive consent from the surface owner to mine the underlying federal coal, other coal companies will obviously be leery of acquiring a lease to this coal. They are certainly not likely to bid much, lacking any assurance that they can actually develop it. The only satisfactory way to deal with the problem is to insist that consent be granted in advance of a lease sale, and that it be transferable automatically to the winning bidder for federal coal on prespecified terms. Negotiating consent and financial terms with numerous surface owners is likely to tax the government's administrative resources. The grant of consent is potentially an extremely valuable right to surface owners and any transfer of this right could well require protracted negotiations in a high stakes atmosphere, hardly the kind of task for which the federal bureaucracy is well suited. Yet, it is hard to see why private companies should want to spend their time and effort to reach a consent agreement that would be automatically transferable to other companies.

The necessity to obtain surface-owner consent before a lease sale creates major obstacles to effective land-use planning. The pattern of granting of consent by ranchers is likely to be scattered and haphazard, with one rancher going along if the price seems right and his neighbor refusing. Plans to achieve coordinated coal development are likely to take a back seat to the ability simply to obtain consent at all. Aggravating the problem is the fact that surface owners in many cases may not make up their mind about granting consent until very late in the planning process. It is typically difficult to bring tough bargaining to a close before it is pressed up against a rigid deadline. Ranchers thus are likely to continue to seek a higher payment for their consent until they fear that they may be passed over altogether. This point may not be reached until the government is almost ready to lease. Again, the determining factor in coal planning could become simply the ability to obtain rancher consent. A year after passing a bill making land-use planning a basic requirement, Congress enacted legislation likely to frustrate this objective as well.

Surface-owner consent offered a contemporary demonstration of the forces that shaped public land policy in the nineteenth century. In passing the surface-owner consent requirement, Congress appeared to serve a noble goal which many individual congressmen may have applauded—protecting long-time western ranchers from displacement against their will by coal mining. In the nineteenth century most public-land legislation had similarly meritorious aims—aiding the smaller settler, stimulating the planting of trees, draining the swamps, achieving the construction of roads, railroads, and other infrastructure. However, most of these laws never worked as intended. Indeed, most of the public land laws in the nineteenth century became vehicles for extensive manipulation and self-enrichment by various private interests. Although too early to say for sure, surface-owner consent shows considerable likelihood of maintaining this heritage. Without a doubt, it creates a very unwieldy division of property rights which significantly complicates the task of administering other public-land laws.

In the long run the congressional giveaways of the public lands in the nineteenth century probably served the nation's purposes well—if unintentionally—by getting these lands one way or another into productive use in the private sector. Perhaps the surface-owner consent requirement will have similar salutary consequences. It gives the surface owner a share of the property rights to the underlying coal and thus may greatly encourage him to seek its development. In the absence of the requirement for consent, surface owners might have offered long and unyielding opposition to federal coal development. Given the numerous handles generally available for obstructionist purposes in the American legal system, and the likely sympathy of local politicians and judiciary, surface-owner opposition might well have proven effective. Despite the unhappiness of the coal industry, it may have actually benefited from the provision for surface-owner consent in that this provision represented the necessary concession to surface owners in order to buy their agreement for further coal development. In short, as noted above, the surface-owner consent requirement was a logical application of interest-group liberalism—and its problems a commentary on the consequences of interest-group liberalism.

In cases such as surface-owner consent, interest-group liberalism actually shows a tendency to converge with the workings of a planned market. In effectively transferring property rights and thereby offering a powerful market incentive to permit federal coal development, surface-owner consent perhaps should even be viewed as the true use of a market mechanism. Because Congress does not seem to have consciously planned it this way, the result might be said in this instance to have been an unplanned market solution—not so unusual in this regard. It was a fortuitous consequence of the tendency of congressional public-land actions to become vehicles for the profit of particular interests. An economist might argue that this result being perhaps a permanent state of affairs, it is better to minimize the number of politically determined decisions. Profit-seeking behavior is instead better directed through market institutions which take it for granted and are specifically designed to channel it for broader social purposes. Whatever the problems of the old EMARS II, there were still many proponents of the view that much greater use should be made of planned markets.

15. The Rediscovery of the Planned Market

The mainstream of the U.S. economics profession has never become reconciled to interest-group bargaining as the basic mechanism for resource allocation. Aside from inequities caused by wide differences in the political power of interest groups, the process of negotiating agreements is seen as very time consuming and socially unproductive. Moreover, the inevitable compromises are frequently made at the expense of rationally efficient use of resources. Political symbolism and face saving compromises often take precedence over cost effectiveness and maximization of benefits minus costs. As one skeptical government economist assessed the Interior coal-leasing program, it might be a big success in diffusing conflict, but it fared poorly when judged by a standard of efficient allocation of coal resources.[1]

Surprisingly, segments of the environmental movement now were also coming to question interest-group liberalism and to see the market in a new light. A 1978 book published by the Sierra Club, reassessing earlier distrust of market mechanisms, now concluded that "the market will force us to do all the right things, even if for the wrong reasons."[2]

The new attraction of the market to environmentalists was its capacity to bring about radical change quickly. Indeed, a main advantage of the market for classical liberals had always been that they saw it as a ruthless but very efficient engine for achieving social progress. Friedrich Hayek comments that the position of the true free-market proponent (unlike many stand-pat "conservatives" whom he disparages) "is based on courage and confidence, on a preparedness to let change run its course even if we cannot predict where it will lead. . . . Especially in the economic field, the self-regulating forces of the market will somehow bring about the required adjustments to new conditions, although no one can foretell how they will do this in a particular instance."[3] Environmentalists were far less sanguine about the likelihood of achieving true progress; however, they didn't doubt the need for speedy social change in order to avert decline or even disaster. For example, many environmentalists favored massive conversion to solar energy and other technologies of slower moving and much more decentralized living arrangements.

Environmentalists recognized that the process of political compromise among interest groups was very often inherently conservative in the sense of holding back change. If society were going to move to a fundamentally new organization any time soon, some environmentalists now reasoned, the market just might be the best way to get there:

> The operation of the market system, whatever its disadvantages, is still the only feasible mechanism to bring about the transition from an affluent society

> to a frugal one. The market will keep us honest, whether we like it or not, since it will be through the market that resource scarcity will make itself felt. The market will be the main force that keeps the consumer from squandering scarce resources, that forces producers to use labor rather than machines, and that keeps the government on the straight and narrow path between depression and runaway inflation. The market mechanism will weaken the bigness of American manufacturing and permit more individual- or family-sized enterprise. It will lead to the decentralization of the economy, to smaller scale technologies, to the repopulating of rural areas, and to reducing the overloading of our cities. . . . The government could help, but given the nature of American democracy, government actions are more likely to oppose the subsidence rather than embrace it, since that is what consumers want. There will be many efforts to deny the market, for reasons that are both noble and selfish, but any effort that tries to deny the reality of scarcity will eventually fail. The society that grew and prospered under the market system now seems destined to subside under the same system. Until the inevitability of finite resources is accepted, there is likely to be a great deal of political bitterness and scapegoating. The peaking of affluence and the initial decline will be the hardest time; we are probably entering this period now. This is the time when the market's role will be the most helpful, almost as a *deus ex machina*. It will be easier and healthier to blame impersonal market forces rather than any of the tradional scapegoats: capitalists, politicians, bureaucrats, labor unions, blacks, Jews, welfare cheaters. And the market will tell us how fast the subsidence must go, by the rate at which raw material prices go up and work opportunities change.[4]

Such environmentalists did not consider private property rights sacred; they had none of the reverence for the market of the old classical liberal true believer. Nevertheless, these environmentalists could now enter a surprising new alliance with older style market proponents. By now there were few true classical liberals to support a genuine revival of the free market—although there was still widespread use of such rhetoric. Rather, market proponents now mainly advocated planned-market liberalism. In dealing with several important issues in 1979 and 1980, the Interior Department shifted back towards reliance on a properly planned market.

Determination of Maximum Economic Recovery

The 1976 Federal Coal Leasing Amendments Act instructs the Interior Department to control the total amount of coal to be mined from a federal lease. Specifically, the Act directs that "the Secretary shall evaluate and compare the effects of recovering coal by deep mining, by surface mining, and by any other method to determine which method or methods or sequence of methods achieve the maximum economic recovery of the coal within the proposed leasing tract. . . . No mining operating plan shall be approved which is not found to achieve the maximum economic recovery of the coal within the tract."[5]

Besides a general conservationist inclination to assert government control in such matters, there were several possible reasons for such a requirement. First, the imposition of the stiff royalty of 12.5 percent created the possibility that coal companies might "high-grade" coal and take only the coal cheapest to mine, leaving all the rest, including some coal worth more than its actual mining cost —and which thus should be mined by an efficiency standard. An inadequate coal recovery could also result from differences between public and private rates of discount of future revenue and costs. If coal companies valued immediate income more highly, they might mine less coal and take a higher annual income for a shorter mine life than was socially preferable. A third reason for the government to intervene to mandate a higher level of coal recovery would be to limit environmental damages. The recovery of more coal per acre would reduce the total number of acres disturbed by surface mining. Coal companies would not have the full social incentive to achieve proper recovery levels because they would not bear all of the environmental costs of surface disruption.

In requiring "maximum economic recovery," Congress did not specify in the law itself just what it had in mind. The legislative history gave some indications that in fact Congress might have wanted to limit the acreage disturbed by surface mining by raising recovery rates above what private companies would undertake on their own. Montana Senator Lee Metcalf commented on the Senate floor that "coal recovery by underground mining methods and by strip mining methods involves quite different impacts upon the environment, upon surrounding communities, and upon land use patterns in affected areas, and . . . [the legislation] would require the Secretary to evaluate and compare the effects on coal recovery by both traditional and novel mining techniques before issuing a lease."[6]

Environmental organizations put forth the alternatives of banning or severely limiting surface mining in the West. NRDC wrote to the Secretary of the Interior requesting "an analysis whether necessary federal coal leasing could and should be confined to tracts that are mineable by deep rather than surface mining methods."[7] A requirement for a greater than private economic rate of coal recovery would be a step in this direction. It would "force more coal to be mined on a given lease, thereby delaying (a) the lateral extension of the land disturbance impacts from surface mining and (b) the beginning of mining in new regions."[8]

An issue paper on maximum economic recovery was presented to the Secretary of the Interior in June 1978 containing five options—only three of which, however, received much consideration. The first option required mining of only those coal seams for which the revenues exceeded the mining costs; this is basically what a private company would mine following its own incentives. The second option ignored individual coal-seam costs and required the maximum total coal recovery possible without making the whole coal-mining operation unprofitable: "extraction of all surface and/or subsurface coal beds within the lease unit displaying the sum of individual present values which is greater than, or equal to, zero."[9] The third option resembled the first, except that it would require a higher degree of coal recovery in those cases where nonmarket environmental benefits were high enough to cover the losses in mining privately unprofitable seams.

The third option offered the ideal solution from a benefit-cost standpoint but posed major hurdles in administration. How could the social benefits of reduced environmental disruption be valued? The prospect of such valuation raised all the old problems of environmental evaluation including the objection that the environment was priceless. In the end, the choice came down to the first two options.

Decisions are made surprisingly often in government simply by whether they will "look good." The first option looked bad, because it essentially let industry decide how much coal to recover, based on private profit incentives. Formally at least, the new coal program was based instead on the conservationist philosophy of asserting a strong government hand over all matters of federal coal development. In these terms the second option looked better; the government directed industry to do something unprofitable with the purpose of reducing environmental disruption. Symbolically, the "public interest" was placed ahead of the "private interest." Based on little more than such symbolism, the Department chose the second option—the maximum recovery of coal achievable without causing the whole mining operation to become uneconomic and close down. In truth, this interpretation was probably the most reasonable guess as to what the Congress actually had in mind in requiring "maximum economic recovery."

The initial decision had been made in a rush amidst many other decisions on the coal program design. But on later close analysis, the option chosen by the Department was wholly untenable, despite its attractive surface appearance. It would have transferred the gains from mining profitable seams to subsidize the mining of unprofitable seams, thus effectively squandering the social benefit of the first seams. The biggest loser would not have been the private coal company, but the federal government, in terms of reduced lease-sale revenues; the bids for federal coal tracts would be diminished by the extent to which companies were required to mine unprofitable seams. If the decision had been fully carried out, many tracts might have had no value at all to the coal companies and would be unleasable.

When wider realization of some of these consequences began to set in, it was clear to the Department that it would have to back off its ill-conceived decision. The Geological Survey aided this back-peddling by producing a study estimating that the decision would increase mining costs by $113 million per year in 1985 in order to avoid the surface mining of 456 acres. In other words, the cost of each acre of surface disturbance avoided by higher coal recovery would be $248,000, an amount the Geological Survey not surprisingly considered much in excess of the likely benefits.[10] Also blistered by numerous industry criticisms of its announced policy, the Department reversed itself, and in June 1979 declared that it would require recovery only of those seams that were profitable to mine individually. Secretary Andrus later stated, "I chose a new definition for maximum economic recovery which corresponds with the manner in which economic decisions are made by industry and is far less costly to this Department to administer."[11]

The decision did not actually preclude the Interior Department from directing a coal company to mine a seam that the company would prefer to leave in the ground. By using a different discount rate, or due to other differences in calculation, the Department might conclude that the coal revenue exceeded the mining costs for a particular seam, whereas the company calculated otherwise. The whole matter though was in effect thrown back into the lap of the field supervisors of the Geological Survey, where it had stood before and where there would be wide administrative discretion. If the Survey followed past practice, there would not be many attempts to compel greater than privately profitable coal-recovery rates.

The debate in the Interior Department over maximum economic recovery occurred because Congress had left the matter so ambiguous. In fact, the final Interior Department decision rejected what could well have been the actual congressional intent, albeit a wholly unworkable one. As a practical matter this mode of interaction between Congress and the executive branch is surprisingly frequent. Congress has a rough idea that some policy or procedure might be desirable, but lacks the resources and technical wherewithal to determine whether it really is desirable or administratively feasible. Congress therefore goes ahead and creates new authority, but leaves the executive branch sufficient maneuvering room to let this authority go unexercised, should the results seem socially undesirable. The ultimate legislative outcome is a joint product between the legislature and the executive—or else, as has become increasingly common, the courts substitute their views for the executive role.

In the case of maximum economic recovery Congress had enacted a logical extension of the conservationist tenets which formally guided most of its actions in public land management. The Interior Department turned down the opportunity to exercise the newly created authority mainly for practical reasons; a considerable search had revealed no desirable social objective which it would be practical to pursue under the given authority. But the decision also reflected the beginnings of a turn away from the conservationist philosophy and a willingness to look more toward the market mechanism—ironically, the point from which the discredited EMARS II coal program had begun. Proponents of interest-group liberalism could concede that interest-group bargaining seemed an inappropriate way to plan such a technical matter as the number of seams and amount of coal recovered from a particular mine.

Fair Market Value

One of the basic criticisms of federal coal leasing, as undertaken prior to its suspension in 1971, was the low level of bidding competition and the low prices received for federal leases. In its much-publicized study, *Leased and Lost*, the Council on Economic Priorities reported that 59 percent of lease sales had attracted only one bidder and that the average bid was only $3.31 per acre. The study concluded that Interior policies had "clearly kept the price of public coal leases too low."[12] According to one Interior official, "As you look back over the seventies and Congress' efforts to reform the Federal coal leasing system, there were two main factors that I see as driving that change. The first was Congress' concern over the rate the Department was leasing coal, and the second was the concern of whether the government was getting fair value from the minerals."[13]

In putting together the EMARS II leasing program, the Interior Department took a number of steps to meet such criticisms. A special group was formed in the Geological Survey to make fair-market-value calculations. The methods and procedures for making such calculations were reassessed and improved. Interior Secretary Kleppe, in formally announcing the EMARS II program in January 1976, highlighted the firm commitment to "return fair market value to the taxpayer through competitive bid sale of coal leases."[14] On the statutory side the Federal Coal Leasing Amendments Act of 1976 specified that "no bid shall be accepted which is less than the fair market value, as determined by the Secretary, of the coal subject to the lease."[15]

The commonsense idea of fair market value for a public output is to find an equivalent item sold in the private market and then take its price as fair market value. If no direct equivalent to the public output can be found, then a similar item could be used, taking account of the differences through some type of comparability formula. But there were basic problems in applying this approach for federal coal. The large share of western coal owned by the federal government limited the number of private coal sales that could be found, especially in certain key regions such as the Powder River Basin. Financial arrangements on private sales also tended to be influenced by or even directly to follow federal terms, making it even more difficult to find a genuinely independent and comparable private sale. Finally, federal coal was such a large share of western coal that the amount of federal coal leased could very well affect the value of both public and private coal deposits.

The main alternative to finding comparable private sales was to estimate fair market value directly, but problems arose in this approach as well. There was no uniform market for coal; rather the purchase of "coal" really constituted the acquisition of a set of valuable characteristics specific to individual mine sites, including energy, sulfur and ash content, coal moisture levels, transportation costs, mining costs, reclamation difficulties, surface-owner problems, proximity to labor force, local permitting hurdles, and others. Differences in these characteristics among coal sites give rise to economic "rents," much as different locations in an urban area generate varying land rents and values. The market mechanisms for determining rent levels are considerably different from price setting for a single standard product manufactured by a number of companies nationwide.

To suggest the complexities introduced, consider coal producer A negotiating with coal owner B to buy B's coal deposit. Producer A would have some idea of the price for which he could sell B's coal. He would then factor in mining and other costs and come up with an estimate of his potential profits. At this point he would have to ask owner B how much B wanted for his deposit. Given B's offered price, producer A could calculate how much his net return would be—i.e., direct mining profits minus the payment to owner B. Producer A would be interested in actually acquiring the coal if the net return from B's coal exceeded the net return he could earn from any other deposit available to him. Of course, A could counteroffer and find out what B's response would be.

Coal owner B would confront the other side of these calculations. He would similarly estimate mining profits available to producer A with his coal deposit and for any other deposit A might mine. Owner B would know that A could go elsewhere and thus he would be constrained in the amount he could get from A. However, owner B might be able to get a higher amount from some other coal producer. He would offer the coal deposit to producer A for no less than the maximum amount he thought he could get from any other producer.

If enough negotiating took place among all coal producers and all coal deposit owners, an equilibrium would eventually be reached where no potential coal producer could better himself by moving to any other deposit, given its purchase price, and no coal owner could find another producer willing to pay more for his coal. There might also be producers who would end up pushed out of the market altogether, because they were not willing to pay the necessary price anywhere, and some coal owners would retain their deposits to wait for a higher price another day. A set of offers acceptable to all parties—the best anyone could do—would constitute the final balance and true market solution.

In practice, such a solution is never reached; negotiations never occur simultaneously among all coal producers and owners. The purchase prices of coal deposits are instead worked out sequentially; only a few negotiations go on at any given time.

A strictly correct calculation of fair market value for federal coal would have simulated such a market solution. Although a major portion of the coal in a region might be federally owned, the simulation would treat each individual tract of federal coal as though it had its own individual owner who negotiated for the highest price possible.

As in other areas of the coal program, however, information costs and administrative feasibility took precedence over theoretical precision. The Interior Department did not propose to try to simulate a complete market solution based on negotiations among many individual producers and coal owners. A vast amount of information would have been required, almost certainly far more than could be justified for the purposes of determining fair market value—one more instance in which a calculation would show that at some point further economic analysis would itself be uneconomic.

Rather than a comprehensive solution, the Interior Department instead focused its fair market value analyses on individual tracts which it proposed to lease. For each such tract an estimate was made of the selling price the coal would bring—derived from a comparison with other similar coal for which price data could be found. An estimate would also be made of the mining cost of the tract. Using these estimates, net revenues would be projected and then capitalized over expected future years of mining to give the Department's overall estimate of the tract value.

For several years this basic procedure was employed for the few lease sales held to supply small amounts of coal to meet urgent requirements (in "short term" lease sales). However, there was not much confidence in the resulting estimates. A systematic reexamination of fair market value was included in the development of the new coal-leasing program in 1978 and 1979.

One major question concerned the coal price to be employed. The only prices readily observable were current coal prices, whereas future net returns would depend on prices in future years. The Department was concerned that temporary supply bottlenecks due to the long moratorium on federal coal leasing might be creating an artificially high price of coal in the short term, possibly causing fair market value estimates to be overstated. A Department task force, gathered to study fair market value, found that

> there appears to be a significant difference between current long-term-contract price for mined coal and the estimated price at which coal on the best Federal coal lands, which are soon to be made available to the market, could be produced. . . . As Federal coal leases become available in increasing number, the selling price of mined coal should drop as competition for coal production contracts drives higher cost coal from the market. . . . The key issue facing the Department is what Federal coal leasing-pricing policy would be best to follow during this period of predicted decreasing real contracted coal prices during which transition and monopoly rents may possibly exist on Federal coal leases.[16]

The task force was troubled that unduly high current prices for federal coal might themselves hold the future price of coal artificially high. "Once high reservation prices and royalties are used, they may be locked into the system for an indefinitely long period, effectively halting movement towards the equilibrium coal price." Moreover, such higher prices would "eventually show up in consumers' electric bills."[17] This analysis was questionable, at least for bonus payments for leases, because they are a sunk cost that should not affect future coal prices to utilities; such prices should be determined simply by the relation of available coal supply to coal demand. Nevertheless, the concern of the fair market value task force that future coal prices might be held artificially high was a valid one. But the more likely cause would be a failure of the Interior Department to lease enough coal. If it were to miscalculate its fair market value estimates on the high side, because of unduly high assumptions about future coal prices, many federal coal tracts offered for lease might go unsold. By means of this mechanism an unrealistically high expectation of future coal prices might indeed become a self-fulfilling prophecy. The task force did emphasize the importance of leasing ample amounts of coal. If the federal government failed to lease enough coal, because of the very large federal share of western reserves, it could drive up prices significantly (or hold off downward pressures). The federal government would in effect be capturing monopoly revenues by holding back supplies in the fashion of OPEC. "The commonly cited recommendation that the Government should seek to capture the entire economic rent should be qualified in the case of coal leasing. First, the Department has potential monopoly powers; it can control prices for leases by the rate at which it releases coal to the market. Clearly these powers should not be used to maximize rent."[18]

Other coal price problems included the difficulty of assessing the impact of coal quality differences on coal prices from a particular site. On the mining-cost side, the Department was taking into account cost differences among coal deposits, but not transportation cost differentials. As a result, the task force stated that "location rents are generally not reflected in the present method of rent evaluation."[19]

Considering the many uncertain elements, the task force recommended that "generally the model should be run with conservative estimates of input." In particular, conservative estimates of long-run coal prices, probably lower than current prices, should be used. "The price used in the discounted cash flow should be the breakeven price of coal from the marginal coal mine. . . . The task force believes that at present it is highly probable this marginal price would be lower than local prices seen in the present market."[20] For small tracts the problems were so great that the task force recommended simply charging a single minimum standard price for all such tracts.

However, the task force was still not really satisfied with these measures. They were in truth a last resort. The Interior Department had been wrestling with the problems of determining fair market value since the mid-1970s. Each new attempt to tackle the issue, rather than moving closer to a solution, had seemed simply to reveal greater complexities and more formidable, practical hurdles. It simply did not seem possible in the real world to calculate an objective number that would clearly be the one true "fair market value" on which all involved parties could agree.

Much as with the issue of maximum economic recovery, the Interior Department's inability, despite much effort, to find a way to calculate a satisfactory government answer eventually led in another direction—to the market mechanism. Although the Department task force on fair market value did not offer a specific design itself, it recommended that the Department try to find a planned-market solution. The Department should "to the extent possible and in a cost-effective fashion, develop a market place for federal coal leases that attracts more than one bidder per tract. Competition is the surest way to capture producer rents (surplus profits). A rent capture policy without a competitive marketplace would likely conflict with other objectives of the program."[21]

The results of the Interior reassessment of fair market value in 1979 and 1980 were a shift away from conservationism. It was all the more surprising because getting fair value was a main battleground in conservationist history. Indeed, the conservation movement originally attracted much of its public support by its vigorous opposition to widespread fraud and illegality in the disposal of public land in the late nineteenth and early twentieth centuries. Aside from the fact of violating the law itself, the biggest objection was that the public was being denied the true value of the lands disposed. The many complaints in the 1970s about failure to get full value in leasing coal thus followed in an old tradition.

But now, in 1980, the Interior Department was saying that fair market value was actually not so important, that the greater danger was the failure to make enough coal available to the market. It was similar in some ways to the long-run verdict on the public-land disposals of the late nineteenth century. Despite the outcries of Gifford Pinchot and other conservationists against the immorality of speculators and defrauders, most historians have concluded that the basic problem was a set of public-land laws enacted by Congress that were impractical and unworkable in the actual circumstances of the West. In the late 1970s the Interior Department was still finding it difficult to carry out the apparent intent of the public-land laws.

The short shrift given to such a key symbol of conservationism as fair market value was possible only because various outside constituencies now were willing to offer support or at least go along. As likely recommendations of the Interior task force on fair market value were put forward for public comment, there was little opposition to the proposed greater emphasis on leasing more federal coal at the risk of losing some revenue. Outside economists serving as consultants to the task force were worried that the Department would try to get "too high prices" which could "lead to zero purchase, zero production and zero rent receipts" of federal coal. As a result, there should be a "bias . . . towards the lowest reasonable number" for fair market value.[22] Economists had long been very critical of the conservationist direction the coal program had taken in recent years. One economics professor serving as a consultant to the task force on fair market value offered the blunt view that "while many aspects of policy could be criticized, my favorite targets are the combination of limiting leasing to that 'needed' for future production, due diligence, maximum economic recovery, and surface owner protection. I consider each idea asinine and their combination the source of intolerable barriers to a socially optimable coal leasing policy."[23]

Another economic consultant to the task force recommended at a public hearing that "the prime objective, in fact, when the government owns property ought to be how to facilitate market solutions to the leasing problem." But a proper market solution would not be achieved without conscious design; it was necessary to plan market features carefully to achieve the desired social objectives.

> There has been a shift in attitude away from considering the idea that if things aren't working quite right, let's create a law. We have discovered that that frequently doesn't work and is counterproductive. Indeed, it usually makes things worse. But there are alternatives in the sense that we are now beginning to recognize that markets are not static things and not simple-minded things like going to the grocery store. They are widely varied. They can be specialized to do special tasks. . . . We need to clarify what is the special form of the market that is needed in this area in order to achieve our purposes and so that individuals, in maximizing their own personal welfare, will unknowingly thereby generate an increase in the social welfare. That is the whole point of the nature of the competitive system.
>
> The wonderfully imaginative and closely adaptive procedure generated by the market is absolutely astounding to those who have no clear conception of how the markets actually operate, and recognition, in fact, that there is a whole variety of alternative procedures which market mechanisms generate to tackle and solve specific problems of the transfer of property or the making of exchanges.[24]

Some environmentalists now publicly supported such views. At the fair-market-value hearings Laurence Moss, a former national president of the Sierra Club, said that he preferred "market signals to tell me whether the leasing rate is too fast or too slow." He generally recommended that the Interior Department should "try to use the market mechanism to the fullest, both in determining the fair market value or the reservation price and in providing information on the desired rates of leasing."[25]

Stimulating Greater Bidding Competition

Exact estimates of fair market value would not be important if the government could simply generate enough bidding competition. In sales of oil and gas leases on the Outer Continental Shelf, for example, the Interior Department has traditionally relied heavily on such competition to assure that it is receiving fair market value. There were some major obstacles, however, to adequate bidding competition for federal coal. Large amounts of federal coal are located in old railroad checkerboard areas where federal sections alternate with sections of private coal. In other areas federal and private coal is intermingled in less symmetrical ways. Wherever both private and federal coal would be needed for a mining unit of economic size, the owner(s) of the adjoining private coal would have a major competitive advantage over any bidder for the federal coal. A 1975 report done for the Interior Department concluded that "the checkerboard pattern of ownership may constitute a significant impediment to efficient mining of coal in the checkerboard, reduce the value of any federal lease to a potential bidder, and reduce competition as well."[26]

As discussed earlier, the frequent separation of surface and coal ownership creates similar formidable obstacles to competition. Given the law's requirement for surface-owner consent, it could turn out that the one company already holding a consent would be the only realistic bidder. Additional factors might restrict competition for federal leases. There may be difficulties of gaining general access to the lease site that favor one coal company over another. One company may have key geologic data that is unavailable to others. The coal deposit offered for lease may be of a size and quality meeting the specific requirements of one coal company much better than others.

The Interior Department had available a number of innovative techniques to try to spur competition for federal leases. One small step would be to guarantee access to the lease site. A closely related, if more complicated procedure, would be to require a transferable surface-owner consent before offering any federal lease for sale. The Department said in 1979 that it would not lease federal coal where consent had already been granted to one coal company, unless the consent were made automatically transferable to any other company making a winning bid for the federal coal. Where the subsurface coal ownership itself is fragmented, the Department has proposed formation of joint leasing arrangements between the federal government and private coal owners. Under such "unitization" schemes any bonus bids (and possible royalties as well) would be divided according to a sharing formula. Bidders would in effect bid for an entire economic unit that had already been assembled containing both federal and nonfederal coal. A 1980 study by the Environmental Law Institute recommended that "in areas of scattered ownership where many private holders are potential bidders for small federal leases, we believe that careful packaging of small federal properties into larger . . . units can enhance the level of competition and result in greater rent retention by the federal government."[27]

The Interior Department has also examined new bidding systems that would promote greater competition. The one which has received the most attention is called "intertract competition."[28] As the name suggests, it would work by forcing bidders to compete not only among themselves for a single tract but with bidders for different tracts. Bidding competition across tracts would be feasible where there were more tracts available for offer than the federal government actually wanted to lease—a common circumstance in areas of preponderantly federal coal ownership. The high bids received for each federal tract offered in an intertract bidding sale would be put on a per-ton basis and matched against one another. The highest bid per ton for any of the tracts would be accepted first, the next highest bid for any tract second, and so forth, continuing to accept lower bids until the overall coal reserve target for the lease sale had been reached. The high bids still not accepted at that point would be rejected and the corresponding tracts not leased at all.

In addition to promoting competition where it otherwise might be weak, this form of bidding would offer a useful means of deciding which tracts should actually be leased. By selecting tracts according to the bid received, the more valuable federal coal deposits would automatically be selected ahead of less valuable deposits. This desirable result would be achieved at a minimum of information gathering and general decision-making expense to the government. Instead, it would maximize the use of industry knowledge about relative tract value. A more complicated ranking procedure would first adjust industry bids by an environmental damage premium in order to achieve a ranking of tracts based on an overall balance of development and environmental values. Intertract bidding would keep the Interior Department out of heated controversies over decisions to offer one tract but not another. Coal companies sometimes have large stakes in such a decision, but there may be very little objective reason for the government to prefer leasing one tract to another.

The proposal to lease federal coal through intertract competition sought to create one particular form of planned market. The idea of intertract bidding was proposed as a possibility in 1975 by the designers of EMARS II in the Office of Policy Analysis, and was kept alive through the next five years partly by this Office's continuing efforts to see it tried out—a limited surviving element of its earlier broad advocacy of planned-market approaches. A number of other innovative coal-bidding techniques and new market mechanisms have also been proposed, but none has received a great deal of attention from the Interior Department.[29] One consequence of the sharp break from EMARS II in 1977 was that Interior study of planned-market mechanisms for federal coal leasing almost ceased for the next few years.

Outside the Interior Department, little was going on as well. Although economists are often quick to recommend planned markets in general, they are not as ready to do the spadework required for the actual design of such markets. During the 1970s the economics literature concerning western and federal coal remained very thin.[30] There was not much available from outside published sources on which Interior coal-program designers could draw. Yet, the federal coal program offered an almost precedented opportunity for market planning by the government for a major U.S. industry.

VI. A Socialist Experiment in America

Introduction

The public lands are an anomaly in the American system in that the federal government directly owns a key means of production. As examined in previous chapters, this ownership gives the federal government the ability to determine production methods, output levels, and other matters normally decided in the U.S. economy by the private sector. Admittedly, the federal government could avoid such tasks by transfering coal resources to private lessees with few restrictions, or by directly selling the coal. However, the Interior Department has imposed diligent development requirements, prepared production goals, and taken various other measures to tightly control federal coal production. Federal coal management, as indeed all of public-land management, is a limited socialist experiment in America. This experiment offers a special opportunity to observe the several ideologies that compete to influence American government. The more normal context in which these ideologies operate is government regulation of private behavior.

The issues raised by the deregulation movement in recent years are very similar to the issues of public-land management. Similar criticisms of public land and regulatory institutions have been made—for example, the tendency for government decision making to become a balancing of interest-group pressures in which economic efficiency inevitably takes a back seat. A widely held theory asserts a strong tendency for government regulatory agencies to be "captured" by one particularly powerful and directly affected interest group—usually the industry that is subject to the regulations. The same critique is offered of public grazing, timber, and other land management. Economists have also proposed various planned-market approaches to substitute for the prevailing direct-command methods of government regulation, much as they have done with respect to the public lands.

The difference between government regulation of private business and direct government management of publicly owned resources can be overstated. Property is often said to consist of a bundle of rights. When government regulates, in effect, it directly controls some "sticks" in the full property right bundle and the private sector has the other sticks. Much the same is true when government leases its coal or other resources for private production; some matters are transferred to the discretion of the private lessee and others remain public decisions. In this light, even government regulation of private business might thus be described as a form of socialism with respect to the set of "nationalized" sticks in the overall property right bundle.

The fact that public-land management in practice does not look much like the traditional aspirations of socialism should not be surprising. Socialism has in many respects proved a much more conservative than radical influence—for example, in art, literature, and morals. Richard Hofstadter argued that, despite a faith in human progress, progressivism in America was actually driven by the desire to preserve old values and institutions in the face of powerful economic and technological pressures for change. In the Soviet Union a revolution in the name of socialism ended by preserving many of the governing and cultural institutions of Czarist Russia in a new form. If the twentieth century has taught anything, it is that radical movements with utopian visions are at least as likely to achieve opposite results as they are their professed goals.

16. Lessons in Political Economy

The Free Market

The nineteenth-century philosophy to dispose of the public lands was well suited to a society which prescribed a minimal role for government and relied primarily on the free market in economic affairs. Yet, very early there were signs of a lack of confidence in the independent capacity for survival of the unrestricted market. Almost all the public-land laws of the nineteenth century placed tight limits on the size of land disposals in order to aid the small settler. Apparently, it was considered that at least in the land market oligopolistic and monopolistic forces might well win out in a free competition. Indeed, in the decades following the Civil War competitive markets of small firms did fail widely by a valid "market" test; in competition with large new corporate organizations, the market of small firms tended to disappear.[1]

The nineteenth-century policy of disposing of federal coal also did not survive long into the twentieth century. Indeed, there were common forces at work. Resource allocation by market interactions among independently owned firms was losing out to the internal resource allocation of the large corporation—in a word, to planning. In the private business world, Frederick Taylor attracted international attention and acclaim as he developed methods and advocated use of scientific management of production processes. Large organizations proved better suited to take advantage of rapid scientific and technological advance and to apply systematic methods in financial, management, and other business areas. As a leading student of the modern corporation has commented, its success was mainly a matter of superior brain power:

> If we know one thing it is that increased productivity, in a modern economy . . . is always the result of doing away with muscle effort, of substituting something else for the laborer. One of these substitutes is, of course, capital equipment, that is, mechanical energy.
>
> At least as important . . . is the increase in productivity achieved by replacing manual labor, whether skilled or unskilled, by educated, analytical theoretical personnel—the replacement of "labor" by managers, technicians and professionals, the substitution of "planning" for "working.". . . The basic factor in an economy's development must be the rate of "brain formation," the rate at which a country produces people with imagination and vision, education, theoretical and analytical skill.[2]

As Samuel Hays has emphasized, conservationism embodies a similar outlook; it sought to apply scientific and expert knowledge to the particular area of natural resources.[3] Gifford Pinchot was in essence the Frederick Taylor of public resource management. Moreover, if large organizations were the most efficient, then it no longer made sense to dispose of public land and other resources. What purpose was there in transfering such resources from one large organization, the government, to another large organization, the private corporation? The government was accountable to the public; it was often unclear to whom the large corporation was accountable. Clearly, corporations were affected by market forces, but also clearly not in the manner prescribed by classical economic theory. Why should government not itself apply scientific management to its own lands and properties?

But the management of federal coal proved hardly scientific. As examined in chapter 2, the Mineral Leasing Act of 1920 actually served to restore the old disposal policy for federal coal, despite its opposite intent. From its enactment until 1971, federal coal effectively was available at low prices virtually anywhere, at any time, to anyone, and could be held as long as desired.

One key to this development was the minimal demand for federal coal. It was also an example of the workings of the American political system under conservationism and interest-group liberalism. Despite a strong professed commitment to planning and scientific methods, conservationism was at least as much a moral and religious movement—proselytizing a "gospel of efficiency." When the evangelical hold of conservationism waned, there was little left to sustain its influence in day-to-day management of federal coal, leaving the field to special interests. Under interest-group liberalism a basic requirement for strong government action is that there must be some powerful group pressing for it. With railroads and a few coal companies the only parties showing any significant interest in federal coal until the late 1960s, a disposal policy had strong support, and in the absence of any opposing interests, it prevailed. The government simply never focused on federal coal matters, reflecting the absence of any visible controversy to compel attention.

Conservationism

The basic progressive and conservationist vision contains many features which in Europe took a socialist form. The same fundamental goal is found—rapid social progress under the banner of science towards a conflict-free, materially abundant future society. The American "scientific public administration" is a home-grown version of European "scientific socialism." Both had their intellectual heyday from around 1890 to 1914. However, reflecting the lack of an American history of monarchy and strong authority, traditional democratic institutions had to be treated more gently in the United States. Thus, Americans would not accept an all-encompassing science of society which would sweep away established political as well as economic institutions. As a result, under the American theory of public administration democratic politics still set the broad objectives, and only afterwards would science take over to carry out the administration of government.

The vision of the promised "good society" which underlay the progressive gospel belonged to the general species of socialist utopias. As Dwight Waldo described this progressive ideal society:

> Their Good Society is strikingly like a World of Tomorrow futurama or Megapolis' fifty-year plan in scale model. Here at last man has become captain of his destiny and has builded [sic] a civilization commensurate with the needs and aspirations of the human frame. It is a civilization primarily industrial and urban—it could hardly be otherwise for "city" and "civilization" are related logically as well as etymologically, and the maintenance of a city nowadays requires industry. It is, of course, a mechanical civilization, for it is the machine that has enabled man to lift himself above his environment and to extend the blessings of civilization to all the members of society for the first time in history. It is quite obviously a "planned" society; such magnificent zoning, for example, would require great imagination in conception and thorough effort and strict obedience in execution. About all we can tell about the form of government must follow from the obvious fact of the planning: it may be "democratic," but the range of government control is unquestionably large and the machinery of administration extensive. It is very probably a "collectivist" society so far as its use of *means* is concerned.[4]

Socialism has shown a great power to inspire passionate commitment of an old-style moral and religious nature; it offered a system of values, a cosmology of society, an inspirational source of personal and group motivation, a prophetic guide to the future—in short, a way of life. But as a system of economic logic, old-style socialism got a failing grade. When socialists first came into power in a few countries, they ran into great difficulties in managing the economy. The experience was bound to force a major reconsideration of socialist economic theory. Socialists began to consider questions such as the necessity of nationalization, limits to central planning, desirable uses of decentralization, and the advantages even of market mechanisms. Oscar Lange and others explored the concept of "market socialism."[5] In western Europe these trends led eventually to the modern social democratic movement.[6]

In the United States the 1970s attempt to finally put the conservationist gospel into practice in the management of federal coal can be compared to the rude shock of the first efforts of economic planning in newly socialist countries. Indeed, in countries that have stubbornly persisted with full-fledged socialist experiments, the early failings of socialism are still seen.[7]

Socialist theory prescribes that scientific analysis should be the basis for economic planning. Thus, in socialist economies,

> plans are intended to embody not the subjective decisions of this or that official or organization, but a scientific analysis of the problems confronting society. Hence, an important role in the planning process is played by scientific organizations. For example, consumption planning is partly based on scientific consumption norms worked out by the relevant organizations. . . .

> To improve the planning techniques there is a continuous process of cooperation between scientific research institutes and the planning organizations. When plans are drawn up, numerous scientific research organizations take part on a consultancy basis.[8]

Federal coal planners are not the only ones who have had to learn that the power of scientific knowledge has often been overstated in the social domain. In his often sympathetic study of socialist economic systems, economist Michael Ellman also finds that socialist planning has not been able to keep up with the complexity of the real world: "The fundamental cause of the waste and inefficiencies described . . . is a theoretical one, namely the inadequate nature of the theory of decisionmaking implicit in the Marxist-Leninist theory of planning." This theory is "inadequate because it ignores the fundamental factors of partial ignorance, inadequate techniques for data processing and complexity."[9]

The U.S. Department of Energy ran into great difficulty in trying to make reasonable coal-production forecasts or set production goals at a high level of disaggregation. Similar problems were encountered in estimating with any precision the consequences of a failure to lease more federal coal. Indeed, the information and administrative costs required to calculate the national pattern of supply-demand response to federal coal unavailability would be huge. If Illinois utilities could not get Wyoming Powder River coal, would they turn more to nuclear power, buy Kentucky coal, or buy local Illinois high-sulfur coal and scrub it? Other possibilities would be to bring in more power from neighboring power pools, or to raise power prices and induce less customer use of electricity? In practice, they might adopt all these measures, some of them, or none of them. How could the federal government obtain good answers to such questions, especially when Illinois utilities do not presently know them? The behavior of Illinois utilities is only one of possibly thousands of such questions involved in assessing accurately the full impacts of federal coal unavailability. To compound such difficulties, the answers to many of the questions can change rapidly, as in fact happened when many key DOE assumptions were rapidly outdated.

The DOE forecasters might want to commiserate with some of their counterparts in socialist countries whose problems sound very similar:

> The problem of slack plans arises from the fact that the necessary information is largely concentrated in the hands of the periphery, and the data available to the centre is heavily dependent on the data transmitted by the periphery. . . . It derives its importance from the inability of the central authorities to concentrate in their hands all the information necessary for the calculation of efficient plans and the complexity of the decision-making process. . . .
>
> [Socialist planning] fails to take any account of ignorance, despite its fundamental importance. It also fails to take account of stochastic, as opposed to deterministic, processes. It assumes a perfect knowledge, deterministic

> world, in which unique perfect plans can be drawn up for the present and the future. In fact, we live in a world in which we are partially ignorant about both the present and the future, and in which stochastic processes are important, and our theories, institutions and policies must take account of this.[10]

In the federal coal program the attempt to implement unsuitability criteria ran up against the hurdle of the very high information cost required to make comprehensive unsuitability determinations. Similarly, although DOE spent many millions of dollars on energy modeling and forecasting, the funds available to such activities could not be increased without limit. U.S. advocates of ever more comprehensive planning often appear to assume implicitly that information and analysis are costless. But, of course, this is not the case; they are often expensive, both in financial resources directly required and in the time and effort of agency personnel. Socialist planners have had to learn these lessons as well, frequently at high cost:

> Not only are the central decision makers unavoidably partially ignorant, but the attempt to concentrate all relevant information in their hands is costly. It is costly in two ways. First, large numbers of people and considerable specialised equipment are required. Secondly, the erroneous view that social rationality can be attained by calculating a central plan which is then faithfully executed may reduce the responsiveness of the country to new information and hence generate waste.[11]

In brief, the experiment in federal coal management conveys much the same lessons for American government planning that have been learned in larger and far more costly experiments in fully socialist countries. Indeed, perhaps the failures of American planning for federal coal can even be turned into a net benefit in the final accounting. Abstract economic reasoning often seems ineffective as an antidote to the appeal of central planning. Perhaps a few small instances of the concrete application of central planning theories in America are desirable as an inexpensive demonstration of the real world obstacles and consequences.

The American progressive variation on the basic socialist vision still closely reflects the basic values held by most Americans. Whatever government really is like, Americans would prefer that it be an objective search for the public interest. However poorly technical expertise typically fares in practice in providing definitive answers to social problems, Americans would still like to believe that there is a valid scientific answer to almost every problem. No matter how heavily government decisions rely on the subjective values, hunches, and intuitions of top officials, Americans want to think that government decisions result from applying the facts of the situation to a scientific principle or other given rule. The great role of the law in American life stems from this same strong impulse to curb any latitude for personal discretion of the administrator. It is because progressive ideals embody so closely the basic beliefs of Americans that progressive influence remains so strong in America—despite a long, historic record of practical failure and a wide rejection among close students of government.

Interest-Group Liberalism

While the management of federal coal provides a good demonstration of the problems of progressivism, how do the alternatives look? What would be the consequences of officially disavowing the existence of "the public interest" and publicly sanctioning an all-out competition for political favor among all interest groups. Indeed, myths and fictions can be essential; it is quite possible that the form of interest-group liberalism observed in practice since the New Deal could not survive official enshrinement as the public philosophy. The widespread beggar-thy-neighbor behavior that would be further encourged would be destructive to all concerned.

Despite the scholarly acceptance of interest-group competition as a valid description of American politics, none of the groups seeking to influence the federal coal program did so with the explicit argument that they had a natural right to enhance their own particular interest. Whatever their true motives, all offered the larger social welfare as the basis for their policy recommendations. The National Coal Association does not advocate greater leasing of federal coal primarily because such leasing would be good for coal companies. Rather, it asserts that the national interest requires more leasing for more coal production. Environmental organizations similarly do not argue for protection measures on the basis that they do not want their old hunting grounds damaged or their favorite view spoiled by strip mining of western coal. Even the proponents of interest-group accommodation as the operating mode of the coal program favor this system because the resulting balancing process is said to promote an equilibrium of interests that in itself is "in the public interest." In short, the current system of interest-group liberalism still functions in the language of progressivism, reflecting the greater appeal of progressive ideology to American values and beliefs.

As a description of events, interest-group liberalism captures only part of the governing process. Obviously, much political activity aims to advance special interests—whatever the arguments formally made. But interest-group liberalism underestimates the critical role of ideology and public philosophy—some would say secular religion. Government is a competition of ideas as well as interests. This book has shown how conservationism, interest-group liberalism, and planned-market liberalism competed to influence the federal coal program. A fourth ideology, evangelical preservationism (descended from John Muir), sought to halt western coal development altogether. A fifth ideology, the old classical liberalism of the free market, influenced the rhetoric considerably but had few real proponents. The program for leasing federal coal was at least as much—probably considerably more—a product of the influence of these ideologies as of the direct special interest pressures.

In examining interest-group pressures on the Interior Department, it is surprising how important symbolic factors often were; in some cases, they were as important as real gains. The coal industry was bitterly disappointed about delays in federal coal leasing that probably would not have made much difference either for total coal production or for profits to individual coal companies. Environmentalists battled for unsuitability criteria and other formal regulations to protect the environment, but which might actually have had only a minor effect on the environmental impact of coal mining, even if they had not been subsequently watered down. Conservationists fought hard for central planning of federal coal, even when they had scarcely any idea how such planning would have to be carried out. The economists in the Interior Office of Policy Analysis lauded the EMARS II coal program as the design of a planned market, even though it was proving very difficult to devise effective mechanisms for such a market.

One of the troublesome questions for free-market proponents has always been the original distribution of property rights. Given such a distribution, the market may simply perpetuate—or worse, aggravate—existing disparities and inequalities in the assignment of rights. A similar question arises under interest-group liberalism with respect to the original creation of interest groups. There is nothing inherently desirable about the existing pattern of groups. J. Q. Wilson, among others, has pointed out that the ability to form and hold together a group may depend upon considerations having very little to do with any intrinsic merits of the group's cause.[12] Indeed, a recent analysis suggested that such problems lie behind current tendencies to rely on litigation to answer so many questions:

> Perhaps the most crucial factor shaping the increased resort to the courts—and the one with the most important long-term consequences—is the growing dissatisfaction and disillusionment with the ability of representative assemblies, at whatever level, to reflect accurately, efficiently, and effectively the desires of the people whom they presume to represent. Over the past decade, public opinion polls have shown a consistent decline in the American public's belief in the efficacy of Congress to solve major problems or protect private rights. . . . In the view of many Americans, their representatives are more the voices of large, organized special interests and less the spokespersons for individual constituents. In short, there is a growing feeling among the public that many of its elected officials and their agents cannot or will not adequately serve the individual interests and needs of the members of society.[13]

In the development of the coal program the most influential group outside the government was the Natural Resources Defense Council, a "public interest" legal organization. In its first few years NRDC had few small contributors and was mostly supported by foundation grants, especially from the Ford Foundation. The staff consisted of a small number of attorneys not long out of the nation's most prestigious law schools. Despite its major impact on coal leasing, it is hard to consider NRDC an interest group. Did it represent the interests of the few staff attorneys? If so, then its great influence on federal leasing was obviously a miscarriage of justice. It is not much of an improvement in this regard to consider NRDC an interest-group extension of the foundations which supported it.

An alternative explanation is to consider NRDC, for practical purposes, as an intelligence-gathering arm of the judiciary. NRDC did in fact perform a function for the courts very similar to the tasks the Interior Office of Policy Analysis performed for the Secretary of the Interior or congressional staffs for congressmen. NRDC gathered coal statistics, reviewed policy studies, proposed policy alternatives, and generally made various analyses for the courts. Under interest-group liberalism the judiciary might be seen as the true interest group. But it would be impossible to justify the individual interests of a few federal judges shaping policies for federal coal leasing.

A more plausible interpretation under interest-group liberalism is that NRDC represented large numbers of people generally concerned about the environment and the impact of federal coal mining on the West, but who had no direct connection with NRDC. Indeed, most of these people never have heard of NRDC. But this may well be stretching the term "interest group" to the point of meaninglessness. In order to preserve the internal logic of interest-group liberalism, any organization with a major influence on government, by definition, must be said to be an interest group.

A much less strained formulation would be to say that NRDC simply functioned as the vigorous proponent of a particular ideology with its own prescription for the public interest. Foundations and other organizations supported NRDC in the belief that its views deserved a hearing in a process of ideological debate; the judiciary sustained many of NRDC's policy recommendations because it agreed with the arguments made and did not shrink from a policy-making role.

The reluctance to accept such a characterization reflects an accurate perception that it leads to some awkward matters. How do we know in a competition among ideologies which one is right? Could it really be arbitrary? Conservationism relied on applied scientific expertise to provide the answer. Finding this wholly unrealistic as a matter of real world experience, interest-group liberalism fell back on an impersonal procedural mechanism of interest-group competition to identify the public interest. To reject both of these approaches can lead to the conclusion that the public interest is a matter mostly of shared values and philosophy without any absolute objective validity. But how could government go about choosing one set of values and philosophy over another, especially when some people may suffer considerable pain? If the government decides democratically, does it simply adopt the values and philosophy of the most numerous part of society or those who are the most outspoken? Or do the values of a special elite dominate? The questions are obviously much easier to raise than the answers are to find. Indeed, much of our current crisis of public philosophy with respect to government reflects the absence of any clear answers.

What was the actual impact of special interest pressures on the federal coal program? It is surprisingly difficult to answer this question, raising issues of true or actual motivation that enter into metaphysical realms. Actions are often described automatically in terms that fit the expectations or ideology of the describer. Thus, a proponent of interest-group liberalism may characterize an action as a response to interest-group pressures, while a conservationist would characterize the same action as the best government estimate of the public interest. The coal program's decision to change the definition of maximum economic recovery is a case in point. It was taken under considerable coal industry pressure to make this redefinition. In fact, the Interior Department encouraged the view that it was bending to industry pressure; under the Department's objective to conciliate all interests, it was desirable that each major interest group should feel that its concerns had received their fair share of acknowledgment in the coal program. But the changed policy on maximum economic recovery was probably made mostly on the basis of Interior's own calculations—to be sure similar to coal industry arguments—that the original policy on maximum economic recovery would cause a huge misallocation of resources and thereby do real damage to the public interest.

Policy analysts in government often first reach their own conclusions as to desirable government policy. For Interior economists the basic policy objective would typically be to promote efficiency in resource allocation from a nationwide perspective. The tactic in advancing such policy measures may then be to look for an interest group which will or at least can be said to offer support. This group may be able to exert significant pressure for the policy. But less commonly recognized, in a climate of prevailing interest-group liberalism, even the legitimacy of a proposal may depend on the existence of outside support from some interest group. Lacking such an interest group, a proposal may have no recognized status.

The impact of interest-group pressures may depend significantly on the type of policy decision being made. The Interior Department comes under the strongest pressures when certain types of actions are involved that have a very concentrated impact. In one case a coal company requested a new "emergency" coal lease to keep an existing Colorado coal mine from running short of coal and perhaps going out of operation. Strong political pressures were exerted by the coal company, employees at the mine, and officials of local government to issue the lease. Letters from congressmen came to the Interior Department requesting lease issuance. The obviously large impact of the federal action on the affected groups, even though they were small, ensured such an outpouring. Partly because there was no similar opposition, the Department in this case liberally interpreted some rules to issue the lease.

But the design of the overall federal coal program was not influenced to nearly the same degree by direct, special interest pressures. In designing the coal program, the impacts of design alternatives on various interest groups often were not clear-cut. No coal company would ever receive or be denied a specific lease because of the program structure. The issues were much more procedural and definitional in the establishment of a coal-leasing framework. If a new U.S. constitutional convention were ever called, narrow interest-group pressures would probably play a much smaller role than in the usual political decision.

Interest-group liberalism sometimes turns out actually to be a mechanism for redefining property rights within a market context. After interest-group bargaining has been completed within the political process, there may well still exist mutually beneficial gains to be made from trade. The competitive market by its very nature exhausts such possibilities, but there is no reason to expect that a political solution will achieve this result. Hence, interest group A might still be willing to make a deal with interest group B for some of B's gains won in politics. Rather than direct barter, cash payment from A to B—or vice versa—is likely to expedite such trading. If this occurs, a market process is created, with B, in essence, selling something to A. The significance of political bargaining would mainly be that it had set the initial entitlements from which A and B then enter into market bargaining. In the coal program this tendency was illustrated by the legal requirement for surface-owner consent and by state coal taxation.

The great appeal of interest-group liberalism is that it avoids questions of substantive government policy. The emphasis is instead on procedure and rules for political participation. In this regard it is much like the old classical liberal ideology of the free market. However, unlike free-market economics, interest-group liberalism has never offered logically well-constructed reasons to show why the results of political competition should result in an efficient use of social resources. Instead, there are a number of logical and practical reasons to doubt that political competition will, in fact, achieve any such happy result. The ends of equity and justice also are not necessarily well served in a system which places the greatest emphasis on procedure and simply reaching agreement.

The strongest case for interest-group liberalism is that it channels human competitive drives that otherwise might break out in far more destructive outlets than political competition. Initially, at least, the ground rules for social interaction must be set by a political process; other than war, there is no other way. Thus, at some level the methods of interest-group liberalism are unavoidable. The question for current political debate is the extent to which political competition should extend beyond social ground rules and into the details of resource allocation. The ideology of planned-market liberalism has represented the chief hope of those who wish to confine the role of political decision making and to leave most resource allocation to the market mechanism.

Planned-Market Liberalism

The current reduced stock of the economics profession is due in no small part to current doubts about the effectiveness of the planned market. The leading model for the planned market, Keynesian macroeconomic management, is severely challenged by government inability to achieve acceptable levels of inflation and unemployment simultaneously. The negative income tax and the pollution tax offer much-discussed planned-market solutions, but so far they have not been implemented. The picture is not wholly one-sided, however. A number of major industrial sectors such as transportation and communications have recently undergone significant deregulation—a turn toward the planned market. Decontrol of oil and gas prices and other planned-market approaches to energy problems have also been adopted.

The obstacles to the planned market have been both technical and political. The initial round of inflation was set off in the mid-1960s by a spurt in expenditures for the Vietnam War without a corresponding rise in taxes—a political decision which economists correctly warned would be inflationary. But in recent years the economy has behaved in ways no economic model could consistently predict; politics aside, the successors to Keynes have been at a loss to know exactly what to do to control inflation without causing widespread unemployment or other unacceptable consequences.

The difficulties which blocked the creation of a planned market in federal coal illustrate the broader problems facing such an approach. In this case, probably the most important social goal that required a realignment of market incentives was environmental protection. Yet efforts to refashion private incentives to account for environmental costs foundered on practical difficulties in calculating the necessary environmental damage charges—much as implementation of air and water pollution taxes has been held back by questions of the appropriate tax magnitudes. The problems were both technical and political. Economists could not provide a definitive technical answer to the value of nonmarket items such as a hunter day or a scenic view at a potential mine site. They could make some good stabs—some admittedly rougher than others—but inevitably there would be considerable disagreement even among economists themselves. On the other hand, without sufficient technical justification, political leaders did not want to take the heat for subjective judgments they might later be forced to defend.

A good case can be made that the planned market was still the best choice, rough stabs and all. The same approximate estimates in any event have to be used later on when a discretionary decision arises as to whether to allow a specific proposed mine. Why not provide coal companies in advance with the government estimate—even if only best guesswork—of environmental costs at each potential mining site? Coal companies could then look for a suitable site to mine that also has low environmental costs; as it is now, companies may select a site and then find that their exploration and design efforts have gone for naught because the government would prefer mining elsewhere for environmental reasons.

The issue in part is whether government or industry should go first. Should industry specify coal mining values at a range of sites, and then have the government factor in environmental costs to reach a final tract selection? Or should government first specify environmental damage charges over a range of sites, and then have industry factor in coal-mining value to reach final mining site decisions? In terms of efficient information management, the party going first should be the one which bears the least cost in making site-specific evaluations. It seems likely that the government could evaluate environmental damages at most sites more quickly and cheaply than industry could determine the coal-mining value of these sites. The argument for government going first is reinforced by the fact that industry is not monolithic; each eligible coal site may absorb the attention and evaluation expenses of several coal companies, but the government alone estimates environmental costs.

Seen in this light, the decision on whether to implement a planned market can be formulated as a management planning question of minimizing total information and decision-making costs to the economy. Proponents of the planned market, however, did not attempt to make such an analysis. The market mechanism is seldom subject to a benefit-cost comparison with other management alternatives; rather, its superiority tends to be assumed, as a matter of faith. In this respect planned-market liberalism itself takes on elements of another "gospel."

In truth, it may be necessary to make many government decisions on the basis of a general guiding philosophy—an ideology—that need not be subject to repeated verification in every individual case. The cost of doing a full detailed analysis of the decision-making costs of the planned market versus the costs of central command decision making by government planners may be too great to justify very often.

The greatest weakness in the case for planned markets is that they themselves require sophisticated government planning. Implementation of a planned market thus encounters the same basic obstacles that have typically frustrated other kinds of government planning. Considerable fine tuning of the market mechanism may be required to achieve given social objectives. However, the normal processes of democratic decision making are unlikely to yield such delicate calculations. In theory, politics can be kept out of the planning of the market mechanism, which would instead be left to economic experts. But such a separation of roles has little more prospect of realization than the similar progressive prescription that administration and planning must be kept strictly separate from politics. The planned-market concept is the best remaining hope for the progressive vision, but perhaps it will prove no more feasible.

Once politics enters, the carefully crafted incentive structure of a well-planned market may quickly crumble. No less in this respect than others, politics is likely to frustrate expert decision making. In federal coal leasing Congress lacked any coherent plan in enacting provisions for diligent development, maximum economic recovery, advance royalties, continuous operations, competitive bidding, and other key instruments of market planning. Market planners in the Interior Department could get around some of the legislative stumbling blocks where there was enough flexibility in the law, but were still prevented from employing various planned-market options.

The old classical liberal ideology was not simply a theory of a market mechanism but a general philosophy of society. The philosophy could not be broken apart and some elements kept while others were discarded as inconvenient. The cornerstone was individual liberty and opportunity, manifested in democratic government on the political side and the free market on the economic side. The political and economic elements complemented one another. Democratic government might tend to frustrate government planning, but it would not make much difference if there were few details to plan, leaving the market free to allocate resources. Moreover, it was not solely a matter of rational analysis but a basic article of faith—a passionate commitment widely held in society—that government planning should be strictly limited to design of the property rights and a few other institutions required for a free market.

A basic problem with the planned market is that it lacks a similar credible political complement to assure that market planning can be accomplished successfully. Its theory of politics is essentially drawn from progressive ideology. The planned market may make very good economic sense, but be of no great help if politics always blocks its successful implementation.

One of the greatest current sources of resistance to the planned market is a matter largely ignored in economics. Westerners resisting market pressures for coal development were distressed most of all by the loss of control over change in their lives. The market is in fact an engine of change; it searches constantly for better goods and services that can be produced more cheaply. This is its great merit, and the source of its efficiency. But all the resulting rapid change is also upsetting to many people; economically speaking, it can greatly affect their "utility." If the rate and manner of change in themselves were considered a basic part of the consumptive output of the economy, a whole new set of scales would have to be applied to measure total economic output. Indeed, many conclusions of economics would have to be rethought. Economic analyses typically avoid this problem by focusing exclusively on the character of a final equilibrium.

As in progressive thinking, planned-market liberalism assumes that society is able first to reach broad agreement on social objectives which then will be achieved by the market mechanism. Proponents of planned markets, however, generally have little to say about how such social objectives are to be determined. Economists typically consider basic objectives to be a matter of "politics," and outside their sphere of concern. This separation of politics and economics is, of course, tenable only if the political determination of social objectives and their economic implementation really constitute distinct realms.

The single most important question of social objectives in the federal coal program was the desirable level of western coal development. However, there was never any explicit political determination made of an appropriate level of such development. The amount of federal coal leased, which would have the greatest federal influence on western coal development, was actually determined as an indirect outcome of an ideological debate over the roles of market forces versus central planning in driving federal coal leasing.

If conservationism is an American version of the old-style socialism of nationalized means of production and direct government command, the planned market may be said to be an American version of a newer, more enlightened form of European socialism. Indeed, many of the more sophisticated socialist economists today are proponents of "market socialism," involving much greater use of decentralization and market mechanisms in socialist economies. In an economic system of market socialism the government would establish independent producers and structure market incentives to achieve social objectives as the decentralized response of producers and consumers to these incentives. This is, of course, the basic idea of a planned market.

American planned markets have usually evolved from free markets; government has imposed a plan on an existing unplanned market. Market socialism, by contrast, evolves from an existing national ownership of the means of production; government then creates new production agents and markets. Although the difference in economic theory may not be so great, the implications for individual liberty and maintenance of political freedoms differ far more.

The Libertarian Possibility

In recent years there has been a growth of interest in libertarian ideas. The liberterian ideology had little influence on the development of the federal coal program. However, the recent history of federal coal management lends support to some of the central arguments made by libertarians. For instance, a recent proposal—libertarian in spirit—to transfer public lands to private ownership reasoned that "so long as the land remains in the public domain, such questions will be decided through political channels. The answer is likely to depend on who can muster the most effective lobbyists rather than on rational calculation. Both parties in the dispute arrogate to themselves the right to define the 'national interest' and to brook no compromise. So long as the issue remains political, it invites confrontation, hostility and denunciation rather than cooperation, negotiation and good will."[14]

Americans tend to regard ideology in a negative way. Nevertheless, the exercise of government power must be guided by some principles and ideas. If there is no credible guide (i.e., ideology) for government action, then it becomes difficult to justify use of government machinery—especially when it involves coercive powers over the citizenry. Put another way, if no public interest is identifiable, either as a matter of actual experience or of theory, then what basis exists for the exercise of government powers?

If not conducted in the service of some guiding principle, government may become simply an efficient instrument to be exploited for private gain. Indeed, the libertarian view is that government becomes a "zero-sum" game in which government powers are used not to create greater national output, but to capture a larger share of a given social pie. Special interests battle among themselves in a beggar-thy-neighbor competition, seeking private profit from the workings of a government lacking any clear purpose. The resources consumed in the competition may even be large enough to shrink significantly the total size of useful social output—thereby creating a "negative sum" game. In short, the old happy view of a balance of competing interests offered by interest-group liberalism is turned on its head.[15]

This book has shown how government sometimes is as much a competition among ideologies as a competition among narrow special interests for direct government benefits. Victory in ideological competition is often defined in symbolic terms; the government takes an action or adopts a policy which represents the triumph of one ideology over another. In the coal program, the policy to base leasing on meeting production goals and the formal adoption of unsuitability criteria symbolized the triumph of conservationism in the early Carter administration. However, the concept that government exists to sanction particular ideological themes, formally acknowledging them by means of symbolic rewards to the political victors, is hardly more appealing than a government which exists to respond to the pressures of special interest groups.

Religious or ideological wars are known as the nastiest of all. There is little basis for compromise when it comes down to a conflict between absolute faiths. Hence, in times of sharp ideological conflict history suggests that it is especially dangerous to make available the powers of government to any party. Rather, the best policy may be to limit the role of government, erecting barriers to the exercise of government powers beyond the basic functions.

It seems likely that proponents of interest-group liberalism have made a critical if mostly unspoken assumption that there already exists a substantial social consensus about the means and ends of government. The basis for such a consensus is that government should behave rationally in the pursuit of traditionally accepted values and ideals. Although critical of the degree of progressive faith in technical expertise, interest-group liberals ultimately also assumed that men of good faith could behave "reasonably" in the common interest. Society would delegate many decisions to professional experts who had responsibility for the application of rational methods in particular fields. Interest-group competition thus would not be too destructive, because it was really a debate about the margins of government action—not the core assumptions and policies.

In fact, this consensus probably existed in American society in the two decades after World War II. It may still exist. However, a number of signs of challenge to old basic assumptions are visible: a new broad public skepticism concerning the capabilities of expert professionals; the doomsday visions and other appeals to obvious religious themes in some segments of environmentalism; fierce opposition to nuclear power as a symbol of the new threats posed by modern science and civilization.

The libertarian movement might itself be added to this list.[16] Its emergence reflects a widening of the field of disagreement concerning government policy, tending thereby to diminish the opportunities for consensus. The true libertarian singles out a particular value—individual political and economic freedom and the absence of government coercion—for preeminence over other social concerns. The use of government coercion becomes for the libertarian what mining in a wilderness is for an environmentalist. In elevating one value to the fundamental standard by which other matters are judged, libertarianism becomes itself a kind of religious force. Ironically, the very fact of this development—by illustrating the breakdown of previous consensus and a growing ideological diversity—adds weight to the libertarian arguments themselves. The greater the extent of social disagreement, the less well government is likely to perform.

However, this assessment extends to libertarian solutions as well. In the absence of a social consensus on a minimum role for government, the obvious question arises as to how government could ever be expected to divest many of its current responsibilities.

This is not the place to examine issues of centralization versus decentralization of government—the appropriate federal structure. However, it might be noted that a growing ideological diversity at the national level provides a good argument for greater decentralization of government responsibilities. Social homogeneity of outlook and thus effective government are more likely to be found in smaller geographic jurisdictions. In this respect libertarian proposals differ significantly from proposals to decentralize government. One proposes to curtail government powers, while the other proposes to maintain government powers by turning them over to smaller units of government where they can more effectively and safely be exercised.

In sum, the libertarian critique of government—much of it by people who would not necessarily characterize their views as libertarian—has helped to expose the weaknesses of the ideologies that justify the current government role.[17] However, the solutions are not as easy; the real world failings of more government also must be assessed against the real world failings of the alternative of less government.

In any case, barring some unforeseen revolution, the future role of government is not likely to be decided in a comprehensive fashion. Rather, it will be decided program by program (or program element) in a highly incremental fashion. The federal coal program will be a small part of this broad trend in American government.

17. The Future of Federal Coal

As the previous chapter emphasizes, there was no one concept or philosophy which fully controlled the development of the program for leasing federal coal. No lasting consensus could be achieved concerning the role of the market, the best type of planning, the appropriate influence of special interests, and other key matters. In such circumstances there is little alternative but to take an incremental approach. By focusing on concrete and limited questions, answers are often easier to obtain. The policies ultimately adopted are likely to appear "pragmatic," i.e., a jumble of solutions to individual problems only loosely linked under some broader umbrella. Incrementalist writings on government most accurately capture the decision-making process as it actually seems to work in normal times.[1]

An incrementalist starts from the current state of affairs and then asks in what ways matters can be improved. The ultimate result is not comprehensively determined, but is the end product of a continual evolution in small steps that may go on for many years. However, an incrementalist philosophy also need not deny the possibility of a more unifying set of beliefs ever emerging. (Indeed, a philosophy which absolutely denied the possibility of any absolute philosophy would be its own contradiction.)

From an incrementalist perspective, the history of the federal coal program during the 1970s can be seen as a drawn-out dispute over whether the initial key step should be taken. The issue that had to be decided first was whether future federal leasing should lend any support at all to western coal development. The bitter opposition to such development had to be dealt with; it simply took ten years to do so.

Incremental Improvements

There are several important areas in which incremental improvements in the federal coal-leasing program could be made. These include the determination of leasing levels, diligent development requirements, and tract selection by means of intertract bidding competition.

The setting of leasing levels should probably be based on an inventory strategy. At any given time there should be a stock of federal coal leases outstanding which contains substantially greater coal reserves than are believed to be needed for new mines to open anytime soon. Any prospective coal mine developer would then have a wide choice from which to select in this pantry of available federal leases. As existing leases were taken off the shelf and committed to production, the federal government would then hold more lease sales and issue further leases to maintain an ample inventory. In this way market signals in the form of existing leases actually entering production would drive new leasing.

This system would encourage and in fact rely heavily on an intermediary role for coal property brokers. Many federal leases would probably not be purchased by their ultimate developer, but by a coal broker who would seek to hold leases until a suitable arrangement could be put together for development. Any limits on the right to resell or transfer federal leases would defeat the whole effort.

The size of the desirable federal lease inventory should be related to time allowed for diligent development of federal leases (and vice versa). Leases should not be issued at a rate so fast as to insure that many of them will have to be canceled later for failure to be developed diligently. For example, a reasonable objective might be to maintain an inventory with five times the amount of coal needed for mines expected to open over the next four years. If this were the case, the diligent development period then should be at least twenty years, assuming a steady growth of coal production.

The ten years currently allowed for diligent development is probably too short a period for other reasons as well. In areas of intermingled federal and nonfederal coal ownership, it does not allow much time for assembly of consolidated mining units; land assembly operations under any circumstances have traditionally been time consuming. Ten years also does not allow much opportunity for negotiations between owners of coal properties and utilities. There would usually be enough time if development of a mine were ready to proceed as soon as the lease were issued. But often this is not the case; time must be provided for land brokering, coal deals to be made, and other institutional arrangements to be put in place. The necessary negotiations cannot really get underway until the availability of federal coal has been assured by leasing it. For a few unusually complex projects such as a gas synthetics plant (which may need coal from a particular federal deposit), ten years may not be enough time even to complete facility construction. A twenty-year period to satisfy the diligent development requirement, corresponding to the current primary term of federal leases, would be a major improvement.

A greater departure would be to allow still longer periods for development of federal leases—thirty years or more. The diligent development requirement is intended to curb speculative holding of federal coal. However, speculation might also be termed "private conservation"—private holding of a resource out of production now because it will be more valuable sometime in the future. By removing any incentive for private conservation by federal coal lessees—indeed prohibiting it—the federal government forces upon itself the burden of determining when federal coal should be produced and how much of it. The diligent development requirement is the key factor driving the federal government to undertake central planning of the western coal industry. This book has shown how such central planning is often beyond the capacity of the federal government. A very significantly relaxed diligent development standard, or even its abolition, would allow the government to rely more heavily on private market incentives to determine the appropriate time for individual deposits of federal coal to be produced.

In determining a suitable inventory of outstanding federal leases, it will be necessary to have some forecasts of the amount of coal likely to be produced in future years. As a result, despite the very approximate nature of even the most skillful production forecasts available, such forecasting should not be abandoned. But there should be much less emphasis on complex computer calculations and more use of surveys of utility and coal company long-run plans. Where computer forecasts are made, the models could very likely be simplified considerably without much if any loss in accuracy. As a consequence, the modeling expense would be less and the ability to understand the logic of model results would be enhanced.

Another area for improvement is in determining just where federal coal leasing should occur. It probably makes sense to establish some nationwide unsuitability criteria. But all coal land should not be checked against each criterion before proposed tracts for leasing have been identified. Such advance checking across the board would be very costly, yet is not essential, so long as the criteria are eventually applied at some point. Nevertheless, where the applicability of a particular criterion can be determined over broad areas without much cost, it may well make sense to do so at an early stage.

The coal industry should be asked to nominate tracts for possible leasing at a point close to if not at the very beginning of the leasing process. If there should turn out to be no coal-industry interest in a particular area, there is little reason to invest government time and effort in checking its suitability for coal development. By focusing analysis on individual nominated tracts, the government can significantly reduce its burden of assessing the environmental consequences of coal mining, as well as other administrative costs. However, the government should not rely solely on industry nominations; for one thing, industry is likely to nominate the least competitive tracts. A company is most likely to go to the trouble of making a nomination where it has a strong reason of its own for wanting that particular tract.

In most cases there will be enough federal coal suitable for leasing that tracts with obvious major environmental problems need not be considered further. Once these tracts have been ruled out, there may well still be quite a few more coal tracts available than the government wishes to lease. Where this proves to be the case, some priorities will have to be set to decide which specific tracts to lease. Probably the fairest and most efficient procedure would be to offer all the tracts for bids and then employ the intertract bidding scheme described in chapter 15. All the federal coal tracts actually receiving coal company bids would be arrayed, starting from the tract with the single highest bid (per ton of reserves) and moving to the tract with the lowest high bid. Going down the list, successively lower high bids would be accepted until the total amount of coal desired had been leased. Bids for the remaining tracts would be rejected. Intertract bidding would also be a useful device for stimulating greater bidding competition and relieving some of the weight from fair market value calculations.

An objection to this procedure is that it does not take full account of environmental consequences. However, it would also be possible to announce before a given lease sale environmental discount factors (cents per ton of reserves) reflecting unmitigated environmental damages at each tract. The discount factor for a given tract would then be subtracted from any bid made for the tract. Hence, the net balance of the company bid minus the environmental discount factor would show the overall coal development and environmental value of the tract—its "overall balance." Conducting an intertract bidding competition with these net values used to set tract leasing priorities would result in the tracts with the highest social value being leased.

This proposal resembles the assessment of environmental damage payments discussed in chapter 6. But coal companies would not actually have to make any payments; their bids would simply be discounted by an administrative accounting factor reflecting environmental consequences. This element, plus the necessity to calculate the factor only for a limited number of tracts, would make such a bidding and ranking procedure more practical.

In any of its versions the use of intertract competition requires that the government set a precise leasing target. Competition among coal companies for tracts then establishes the minimum price necessary for any federal tract to be leased. An alternative basic strategy would be to set an absolute minimum price for federal coal and then lease all federal tracts which received a bid higher than this price. There would be no calculation of fair market value on a tract-by-tract basis. The minimum price would be set high enough that only the higher quality coal tracts would be leased. Poorer tracts could not meet the minimum bid required, thereby preventing long-term speculation in federal coal reserves. By making the minimum price higher or lower, the total size of the inventory of outstanding federal coal leases could be controlled. As it is the case with the leasing-target approach, the minimum price strategy would also be greatly aided by a longer period allowed for diligent development or by the abolition of this requirement.

The issue of a fixed coal quantity (with coal prices set by the market) or a fixed coal price (with coal quantity set by the market) is analogous to the debate over emission fees versus marketable permits in the field of environmental pollution. In each case there is a desired inventory to be achieved—either of coal leases or of pollutants in the air and water—and more than one market mechanism for achieving this inventory.

The surface-owner-consent requirement of the Surface Mining Control and Reclamation Act of 1977 remains one of the biggest headaches for federal coal leasing. This requirement makes it difficult to plan federal lease sales because the grant of surface owner consent may occur—or become known—only very late in the leasing process. The government thus may not know which coal is even available for lease until this stage. The consent requirement also threatens to transfer significant parts of the value of federal coal to the private surface owner. The best and most direct way to resolve such problems would be for Congress to repeal or significantly modify the existing consent requirement. At a minimum, if the current requirement for consent is to be maintained, a legal limit should be established on the amount of payment permitted to the surface owner for his grant of consent. The Interior Department might be able to establish such a limit administratively simply by refusing to lease federal coal where consent payments have exceeded the limit. However, a problem with such an administrative tactic is that it may preclude the leasing of some desirable federal coal deposits where surface owners believe they can hold out for higher payments.

Many federal coal tracts are of interest to only one potential buyer. Such a tract could be adjacent to an existing private operation and be too small to stand alone as a separate coal mine. In the private sector this circumstance would result in a negotiated agreement between the parties. Competitive bidding may make little sense where there is only one potential coal purchaser. Yet current law requires that all coal leases be awarded by competitive bidding. Modification of the law to allow negotiated sales where there is only one prospective user of federal coal would add desirable flexibility.

The royalty rate of 12.5 percent on surface-mined coal is probably excessive. Because federal coal is such a large share of western reserves, this royalty tends to be adopted for private as well as federal coal. One effect of the royalty is to make it unprofitable to mine some coal that could be mined profitably in the absence of the royalty. The main distributional effect of the royalty is to transfer revenues from utilities and other coal users to federal and western state governments—mainly the latter. There is no particular reason for such a revenue transfer; the public infrastructure and other public costs of coal mining could be recovered through more direct state and local taxation of mining. In fact, existing state and local taxes are already ample for this purpose. Montana and Wyoming have had high state severance taxes on coal for some time. The precise coal royalty figure of 12.5 percent was actually selected by the Congress because this amount had long been the minimum permissible royalty rate for federal oil and gas—hardly a sound basis for setting a federal coal royalty.

In areas of highly fragmented coal ownership, the federal government might consider land exchanges or other means of consolidating federal and nonfederal ownerships into more efficient mining units. In the old railroad checkerboard areas such an approach would be particularly desirable. Another approach would be the joint leasing by the federal government and the railroads of consolidated blocks of federal and nonfederal coal large enough for an efficient mine. The two parties would agree to share coal revenues according to a formula established prior to leasing. The result would resemble in some ways the "unitized" development of oil and gas tracts.

In requiring fair market value for federal coal, the government must be careful that it does not behave like a monopolist. The government could earn high coal revenues by withholding federal coal from the market, driving up its price. However, the objective should be to drive the price down to the level that would exist in a truly competitive market for federal coal. With the abundant supplies of federal coal, this price may be quite low, indeed, and perhaps lower than many observers would think is fair market value for the coal.

The government capacity to estimate accurately the fair market value of federal coal by its own presale calculations is very limited. Hence, the emphasis should be placed on stimulating bidding competition for federal coal as the primary means of assuring receipt of fair market value.

Transfer of Federal Coal to the States

A basic alternative to continued federal management of federal coal would be to transfer ownership of the coal to the state in which it is located. Conflicts between coal development and environmental values could then be resolved separately by each state according to the specific preferences of the citizens of the state. The inducement for the state to allow coal development in the national interest would be the revenues generated by coal mining—coming to both the state public and private sectors. Such revenues could be assessed against any adverse impacts on the state of coal mining. Each state, therefore, would be able to decide for itself whether or how much coal revenue should be sacrificed in order to achieve greater environmental protection—either for costly measures to mitigate environmental damages or by excluding coal development altogether. In short, the states themselves would respond to market incentives for coal production to the extent they perceived it to be in their own best interest.

The objection is sure to be raised that transferring federal coal to the states would be "giving away" the federal government's resources. However, in substance if not form, the federal government actually has already transferred much of the rights to federal coal to the states. The states now receive directly 50 percent of the lease sale bonuses, royalties, and other revenues from federal coal. The federal government receives directly only 10 percent. The remaining 40 percent goes into the Reclamation Fund—a special federal fund committed to western water projects. Because this fund mostly replaces direct federal appropriations that would otherwise be necessary, it is probably best considered a revenue source for the federal government. As a result, coal revenues are effectively split 50 percent to the states and 50 percent to the federal government.

However, the effects of corporate income taxes must also be taken into account. Like other expenses, company payments of federal royalties are deductible from corporate income subject to taxation. Since the corporate income tax is 46 percent at the margin, corporate tax payments to the federal government may fall by up to 46 percent of any federal royalties collected, depending on the exact tax status of the corporation. Since the federal government only keeps 50 percent of the royalties (the other half going to the states), federal royalty collections may yield the federal government as little as 4 cents for each dollar of royalty collected. If payments to the Reclamation Fund are not counted as federal revenues, the federal government almost certainly loses substantial net revenue by the imposition of a federal royalty.

Moreover, the costs of coal management have not yet been considered. When these costs are factored in, the fiscal benefits to the national taxpayer (losses) are reduced (increased) still further. In 1978 the federal government spent $39.1 million in coal management and took in only $11.3 million in coal revenues. Despite losing $27.8 million on its coal management, the federal government directly distributed half the coal revenues ($5.6 million) to western states.[2] In the future federal coal revenues will be far larger as federal production accelerates and federal leases are renegotiated to the new royalty of 12.5 percent.

Besides most of the fiscal benefits, the federal government has also already transferred a considerable part of the decision-making responsibility for federal coal to western states. The regional coal teams responsible for managing preparation for coal-lease sales contain two state representatives out of five coal-team members in all. The BLM land-use plans prepared in coal-lease areas are required under the Federal Land Policy and Management Act to be "consistent with State and local plans to the maximum extent . . . consistent with Federal law and the purposes of this Act." As a practical matter, whatever the institutional forms, western states will exert a very substantial influence on the coal-leasing decisions of the Interior Department.

As a result of the financial terms of federal coal ownership and western influence on its management, transfer of federal coal to western states thus might be regarded as little more than bringing the outward form of ownership into accord with its current real status. In this respect, such transfer to the states would not be a radical but really an incremental change. Indeed, the more radical change would be to accord the federal taxpayer the full benefits of federal coal ownership. However, as a matter of ideology and symbolism, the opposite is true; continued federal ownership maintains the form of traditional progressive institutions for the public lands and does not directly challenge conservationist ideology.

One practical objection often raised to state ownership is that the states have been poor resource managers on existing state lands and probably would continue this history for any newly acquired resources. But existing state lands are typically scattered sections left over from old state-grant lands whose geographic pattern would make them almost impossible for anyone to manage. Environmentalists also suggest that state governments would be more subject to special interest pressures and would be more willing to sacrifice the environment for development. But if this were once true, it becomes steadily less so as states such as California, Colorado, Arizona, and New Mexico reach (or exceed) the rest of the nation in degree of urbanization and income levels. In California the state government fully matches the federal government in responding to the demands of its citizens for environmental protection. Environmentalists may be shortsighted in assuming that the federal government will not at some point press for more rapid energy development than the states want.

Indeed, the most serious objection to transfer of federal coal to the states could well be that they would develop this coal too slowly. Moreover, their reason might not be protection of the environment as much as revenue. A few key states, such as Montana and Wyoming, might very well withhold coal supplies in order to push up prices. OPEC has shown how key energy suppliers can cut production and earn larger revenues as a result. A Rand Corporation study some time ago characterized the 30 percent coal severance tax imposed by Montana as "OPEC-like revenue maximization."[3] Transfer of federal coal to the western states might further deepen divisions which already exist among energy-rich and energy-have-not states.

If the surface estate were not transferred to the states as well as the coal, coordination of federal surface with state subsurface management could be a problem. If the federal government does retain the public land surface, it might want to keep the coal as well simply to protect surface uses. In any case, the problem is reduced by the fact that more than 50 percent of the better quality federal coal lies beneath private surface. In addition, the total surface acreage disturbed by coal mining at any one time would not be so great that a significant portion of the public lands would be affected.

A modified version of this alternative would be to transfer most federal coal to state ownership, but to retain the few areas of of highest coal value. The federal government might, for example, retain only its coal within "Known Recoverable Coal Resource Areas," as formally identified by the U.S. Geological Survey. Or it might be sufficient to retain only federal coal in the coal-rich Powder River region, transferring all other federal coal to the states.

Such an approach would effectively treat high-value federal coal areas as economically "critical areas" for the whole nation. The land-use reform movement at the state level has proposed to limit state controls to "critical areas" for the whole state, leaving land-use controls in all other areas to local government. Extending this logic to the national level, the federal government need only become involved in management of those public-land areas which have a critical national significance—economic or environmental. The remaining areas could be turned over to state and local governments, or to the private sector.

The extent of federal involvement in western state affairs as a result of federal land and resource holdings is an anomaly in the U.S. system of federalism. This system is generally based on the concept of maximum decentralization of governing authority, and correspondingly favors centralization only where a clear necessary reason for it exists. For much of the federal coal holdings, no such reason is apparent, other than historical accident and longstanding practice.

On the other hand, an even stronger American tradition is to limit the role of government to those functions which cannot be or are poorly performed by the private sector. Federal coal ownership is a major anomaly in this regard as well. State ownership would simply substitute management by one level of government for another level of government, and thus, quite possibly, one inefficient form of public management for another. This possibility suggests the alternative of outright divestiture of federal coal holdings to the private sector.

Sale to the Private Sector

The central planning of western coal undertaken by the federal government in the late 1970s reflected the fact that federal coal is such a large part of western and U.S. coal holdings. However, if the federal coal were sold off to numerous owners in the private sector, the forces of market competition could take the place of central planning in allocating western coal resources.

The pressures for efficient coal management would be greater in the private sector than they are now in the public sector. Private coal owners would search actively for coal buyers, thereby creating a competition to offer the lowest-cost coal. Episodes such as the ten-year unavailability of almost all unleased federal coal would not occur under private ownership of the coal.

There is agreement across a wide political spectrum that private industry provides more incentives favoring efficient management than does governmental bureaucracy. Arthur Okun, for example, considered that there is a "big trade-off." On the one hand, private enterprise is more efficient; it delivers more goods and services at lower cost. On the other hand, the system of private ownership also tends to promote inequality. Okun gave the efficiency of the mostly private American economic system a high grade: "In comparison with any other production system in man's history, or any blueprint currently on any drawing board, the American economy must get a high performance rating."[4]

Few people have asserted that the purpose of the federal coal program is to promote greater social equality. Indeed, the greatest current beneficiaries of federal coal ownership are a few western states that are already well off due to other large revenues that they earn from natural resources. Wyoming is projected to receive more than $1 billion in total natural resource revenues by 1985—almost $2,000 per capita. In short, if current federal coal management does not serve to promote equality, perhaps greater efficiency should be sought by means of private ownership.

Much of the objection to selling federal coal into private ownership reflects the still strong influence of conservationist ideology. The conservationist conviction that public management is inherently desirable is still widely held; one might say that public management has an important "existence value" for many people as a symbol of the progressive faith in human progress under the banner of technical expertise. However, the actual experience of government bureaucracy offers little support for this conviction. Indeed, almost no one any longer asserts that public management is preferable to private management because it is more efficient. Any claims for public management rest instead on a preference for public over private objectives.

To propose transfer of federal coal to private ownership is not to say that government should leave all coal production matters to the private sector. However, the normal mechanism for dealing with market failings is not direct government ownership of productive resources; rather, it is government regulation—or the design of new property-right institutions to internalize within the market formerly external elements. Under the provisions of the Surface Mining Control and Reclamation Act of 1977, government regulation of surface coal mining will be largely the same whether the coal is federally or privately owned. Moreover, the administrator of the surface mining regulations in either case will be the same state agency—operating under federal guidelines.

Several other objections no doubt would also be raised to sale of federal coal. Government ability to control the impacts of coal development on local and regional land use would be diminished (although there is very little in the past record to suggest that state and local governments are good at either planning or controlling land use). Another concern would be the possibility that coal sold by the federal government would end up in the possession of a few large energy corporations. Fear of concentrated ownership of coal and other energy resources was a key reason mentioned by Congress for the original creation of a leasing system under the Mineral Leasing Act of 1920. It has remained an issue capable of arousing strong popular passions—as shown by the unfavorable image of the big oil companies. To be sure, the existing antitrust laws could be enforced against any excessive concentrations of coal-reserve ownership.

The most telling objection to selling federal coal is that it might be impossible to sell the coal for anything like its actual value to the federal government. Since federal coal holdings far exceed any near-term demands, most coal would have to be sold to long-term speculators. In itself, there is nothing wrong with this; as already noted, speculation is simply private conservation in response to financial incentives. The more relevant point is that the government may prefer to speculate in holding its own coal. The coal may be worth more to the government as a speculator itself than it could bring in from any private speculator.

The worth of an asset held for speculative purposes depends on its eventual future development value and the extent to which that future value is discounted to the present. Assuming that government and private sector perceptions of future development value are similar, the relative valuations of a given asset would depend on their relative discounts applied to future returns. The permanence of the government and its much greater ability to diversify risks make it likely that the government does not discount future revenues to as great an extent as would any private speculator. Hence, federal coal will have a higher "present value" to the government.

Consider the case of oil shale. The current selling price of an oil shale deposit may be low, reflecting the bleak prospects for its near-term development. Facing the possibility that oil shale production may not get underway for many years, a private speculator would not be willing to offer much for federal shale deposits. As a result, the federal government may well maximize its long-run financial return by holding its shale resources until shale development is much closer to occurring. The same considerations would apply to federal coal resources that are still far from the development stage.

One way to reduce the concerns about receiving a fair price for federal coal would be to sell the coal with an overriding federal royalty. The federal government then would still be entitled to significant future royalty revenues whenever development occurred.

A Proposal

There are perhaps 10 million acres of prime coal land owned by the federal government—a total area equal to about 15 percent of the state of Wyoming. More than half of the land containing prime federal coal has nonfederal ownership of the surface. A rough estimate of the total market value of all such coal lands is $5 to $10 billion.

The best way to ensure that this prime federal coal is put to efficient and productive use would be to sell it to the private sector by competitive bidding. The federal government would retain a royalty of perhaps 5 percent—all going to the U.S. Treasury. A restriction might be put on the extent of private acquisition of federal coal lands by any one company—say no more than 10 percent of the total coal acreage could be sold to any one company nationally, and no more than 20 percent to any one company in any single state. The federal coal might be sold off gradually—say over ten to fifteen years—in order to give potential purchasers time to acquire financing and make other preparations.

States would have the option to make up any losses in coal revenues from the federal government by raising (or imposing) their own severance taxes. They could also control and direct coal development by means of the various government regulatory powers already available with respect to private coal development.

Substantial federal coal outside the prime coal lands is not likely to be developed for at least 50 to 100 years—unless further exploration reveals particular deposits to be of a higher quality or more economical to mine than now believed. The coal rights on these lands should be treated in accord with a broader policy for all federal mineral rights. A separation of surface and mineral rights causes many problems; hence, where the surface is privately owned, the mineral rights outside prime coal areas should be offered first to the current owner. Assuming no special known mineral value, the price for mineral rights might be in the range of $50 to $100 per acre. If the existing surface owner were unwilling to pay the government's asking price, the mineral rights would be offered to any taker. If there were more than one interested buyer, the rights would be sold by competitive bidding.

Where federal coal lies outside the prime coal lands beneath federally owned surface, the disposition of the federal mineral rights thus should be tied to the federal policy for the surface.[5] If the surface rights are sold or transferred to the states, the mineral rights should be transferred in a corresponding fashion. If the federal government retains the land, it should retain the mineral rights until such time as mineral development becomes a more realistic possibility. Thus, when development of a federal coal deposit begins to look more promising, the federal government should sell off this deposit, thereby assigning private market mechanisms the function to work out the precise timing and manner of coal development.

Epilogue

On January 13 and 14 of 1981, seven days before the Carter administration came to a close, the Interior Department held its first regular sale of federal coal leases since 1971.[1] The leases sold contained 87.9 million tons and were located in the Green River–Hams Fork coal region in northwest Colorado and southwest Wyoming. Over the next two years regional sales of federal coal leases also were held in the Uinta–Southwestern Utah region, the Appalachian region, and the Powder River region. Overall, these sales resulted in the leasing of 2.0 billion tons of coal, a 12 percent increase in the total amount of federal coal under lease. Further sales have been planned in 1983 and 1984 in these same regions, as well as in the Fort Union region and the San Juan River region. The sales have been planned to offer in the range of 4.0 billion to 8.6 billion tons of additional federal coal for lease.

The Powder River coal lease sale was held on April 28, 1982, and was the biggest sale by far, resulting in the leasing of 1.5 billion tons. As the first sale under the new coal program to offer such large amounts of federal coal for lease, this sale had been intended as the showpiece for the new federal coal program developed by the Carter administration. Planning for the sale dated back to June 1979 when Interior Secretary Andrus announced a preliminary target of 776 million tons for a lease sale to be held in early 1982. This target was more than doubled by the time the sale actually took place. The Powder River coal region supplied 70 percent of the federal coal produced in fiscal 1982 and, as explained previously, is the most important of the coal-producing regions in the West.

The high hopes of the Interior Department for the Powder River lease sale were not to be realized, however. In the same week that I was receiving the page proofs for this book (April 1983), the Interior Subcommittee of the House Appropriations Committee held hearings in which the federal coal-leasing program was the center of attention. The Surveys and Investigations staff of the House Appropriations Committee reported that the Interior Department had mismanaged the Powder River sale and failed to receive fair market value for the leases sold.[2] Coal value data had allegedly been leaked to industry in advance of the sale; moreover, the Interior Department shortly before the sale had changed its bidding procedures in a way that appeared to many to reduce arbitrarily its minimum acceptance price for leases. The criticisms of the congressional investigative report received wide media attention, making many people aware of the large values at stake in federal coal leasing and of the growing importance to the nation of federal coal production.

The House investigative report also questioned the holding of such a large sale of federal coal leases when the coal market was depressed. The reversal in 1982 of the upward momentum of world oil prices, combined with the effects of a general economic recession, had lowered expectations of future coal demands and reduced prices substantially below the levels of a year or two before. Many sellers of privately owned coal reserves were finding it difficult to find buyers or were forced to accept sharply reduced prices.

Two weeks after the report of the House investigative staff was issued, the General Accounting Office also released a report on the Powder River coal lease sale.[3] The GAO report came to similar conclusions, finding that the Interior Department estimates of fair market value for the coal leases sold were about $100 million below the true market value. GAO also recommended that further federal coal leasing be halted until the Interior Department made necessary changes in its procedures to assure future receipt of fair market value.

The Interior Subcommittee of the House Appropriations Committee at this writing (May 1983) is considering a proposal to impose a moratorium on federal coal leasing during the remainder of fiscal 1983, with the possibility of a further moratorium for all of fiscal 1984. As many as six sales of federal coal leases could be delayed or cancelled, involving potential leasing of several billion tons of federal coal.

The Interior Department criticized the quality of the House investigative report on the Powder River sale.[4] The Department's response pointed out that the total high company bids accepted in the sale exceeded by 19 percent Department appraisals of lease value made at the time of the sale. Using later recalculations, the GAO report stated that the Interior Department had received 95 percent of the Department's own field appraisals of tract value.[5] The Interior Department objected that the GAO then developed its own new appraisals—in the Department's view an inappropriate second-guessing of the original Interior estimates.[6]

The House investigative report and the GAO report were both initiated as a result of suggestions of improper industry influence or other improper actions in the Interior Department's conduct of the Powder River lease sale. Although both reports were critical of the Interior Department's conduct of the sale, neither could establish any relationship between this conduct and a loss of fair market value. The conclusions that the Interior Department received less than fair market value were instead obtained by respecifying the technical assumptions and procedures of the appraisal models employed by the Department. Illustrating the hazards of recalculating fair market value, the respecifications made by the House investigative staff differed significantly from the respecifications made by the GAO. Thus, although both reports were recommending corrections in technical appraisal procedures, in significant part they recommended different corrections.

Appraisals of fair market value are highly sensitive to even small changes in the specification of assumptions—all within a plausible range. Moreover, as the owner of 60 percent of western coal, and with indirect control over another 20 percent, the federal government effectively makes the market for western coal reserves. Its leasing policies thus can affect greatly the value of leased reserves, making even the concept of fair market value ambiguous. Indeed, it is more proper to regard the Interior Department as itself setting the market price for western coal reserves. This price should be set to achieve the desired amounts of federal coal reserves under lease and other leasing policy objectives of the government.

For me, the recent episode adds further weight to the remarks made earlier in chapter 15 on fair market value. (These remarks were written two years ago, and I have resisted any inclination to revise them in light of the recent developments.) The problem of fair market value continues to defy government resolution, particularly in the circumstance of dominant federal coal ownership and the absence of federal leasing during the 1970s. The only answer is to accept a somewhat arbitrary figure—as, for example, a 12.5 percent royalty is imposed on all surface-mined tracts—and to seek wherever possible to generate active bidding competition for federal leases. If obtainable, such competition in itself establishes the fair market value of the lease.

Changes in the Coal-Leasing Program under the Reagan Administration

The recent events in connection with the Powder River coal lease sale have jeopardized the plans of the Reagan administration to lease large amounts of federal coal. The method of determining coal-leasing levels was one of several important changes made by the Reagan administration in the Federal Coal Management Program.[7] The Carter administration's system of central planning of coal production goals and leasing targets was quickly jettisoned. The central planning approach was obviously at odds with the Reagan administration's philosophy of using the market. Instead, a new approach—called "leasing to meet the demand for reserves"—was adopted.

The Reagan administration explained that, instead of a primary objective in federal coal leasing to meet a production goal, the new primary objective in coal leasing would be to lease enough coal to create the circumstances of a competitive western coal market. Development of a federal coal deposit would take place only when it obtained a favorable verdict in a market competition with numerous other potential western coal supply sources—federal and nonfederal. Because the federal government owns so much of western coal, and indirectly controls the development of considerable additional coal, a competitive western coal market will come into being only if federal leasing policies are explicitly designed to achieve this objective.

The proposed high levels of federal coal leasing necessary to create such a competitive coal market in the West quickly revived an old leasing controversy. Critics complained that "leasing for reserves rather than leasing to meet a production goal, will result in an over supply of recoverable reserves that will not be developed in the near term." In response, the Interior Department replied that

> the objective of the Department's policy of leasing to meet the demand for reserves is to provide sufficient leased coal to allow for full industry competition for coal supply contracts. In order to achieve this objective, the coal supply under lease must exceed the amount that will be dedicated to development in the near term. By no means should this be considered to be an "over supply" situation. Such a coal supply is necessary for the efficient development and for the proper pricing of the coal which is dedicated to development.[8]

Another widely expressed concern was that the achievement of fair market value would be sacrificed if too much federal coal were leased. Critics suggested that the Department of the Interior "is consciously offering for lease vast areas of federal lands during a time of recession in our Nation which restricts the ability of industry to compete effectively and fully for such coal leases. By flooding the market, DOI is consciously depressing the market and thereby reducing its net return." The Interior Department, however, defended its new coal-leasing strategy as necessary to achieving a competitive market:

> The intent of the Department in providing broad offerings of federal coal leases is not to flood or depress the coal market. Of course, coal leases in a fully competitive market for coal supply contracts will be worth less than coal leases in a restricted market. Thus, the Department can expect to receive lower bids for its leases under a policy of broad offerings than under a policy of tightly restricted offerings. Such pricing is the natural result of the Department's decision not to act as an OPEC-like monopolist, i.e., not to withhold federal coal resources from the market in order to maintain artificially high price levels.[9]

Another change made by the Reagan administration concerned the requirement for diligent development of coal leases. The Federal Coal Leasing Amendments Act of 1976 required development within ten years of all leases issued after passage of the Act. However, the Interior Department was left administrative discretion to define diligent development for leases issued prior to 1976. In 1976 the Department had in fact promulgated regulations requiring development of these "pre-1976" leases by 1986. However, new Interior Department regulations issued in 1982 instead allowed the older leases ten years beyond the date at which the twenty-year initial lease term expired. Hence, instead of 1986, the many coal leases issued in the late 1960s would have until the late 1990s to begin production (i.e., the twenty-year initial lease term plus the ten further years to begin production).

The Interior Department explained that this change in the diligence requirement was being made partly on legal grounds; the Department might not have the authority to impose a unilateral change midway through the initial lease term. A looser diligence requirement also met the policy objective of the Reagan administration to rely on market competition. As the Department noted,

> For most leases issued prior to August 4, 1976, the proposed regulation would spread the production from individual leases over a longer time horizon. This is expected to result in more efficient, lower cost production because, unlike the current rule, it would not induce premature production. This should reduce costs and prices to industry and consumers relative to those that would result under the existing regulation.
>
> The proposed regulations should also have a positive effect on competition, employment, investment, innovation, and on the ability of domestic firms to compete with foreign firms in domestic and export markets. This is likely to be true because the time extension afforded by the proposed rule would maintain a ready supply of leases that provide the most efficient supply for future coal users.[10]

An effect of relaxing the diligence requirement for old leases, however, was to undercut one of the main reasons in the past for leasing new federal coal. During the Carter administration, the need for new leasing had been justified by the assumption of a 1986 production deadline for the large inventory of federal reserves leased prior to 1971. Hence, any new mines formed after 1986 would clearly have to have newly leased federal reserves. But, now, if sufficient old leases were to be left available, new leasing would have to be justified by the relatively superior economic and environmental characteristics of the newly leased reserves—a considerably more difficult case to prove to skeptics. The Interior Department would now have to rest its full leasing case on the policy objective to put large amounts of federal coal under lease in order to create a competitive western coal market. The events of the Powder River sale have complicated an already difficult task of persuading actors in the political process to accept the economic merits of this case.

Two other important changes in the coal program were made by the Reagan administration. Rather than wait until a later planning stage, industry would be invited early in the land-use planning process to identify specific coal tracts for possible leasing. In this way, the application of unsuitability criteria could be focused on those coal areas with the greatest industry interest and the greatest potential for coal mining.

The other important change involved the role of the regional coal teams and the procedures for federal consultation with state governments. In the new regulations for the coal-leasing program issued in July 1982, the role of the regional coal teams was modified to provide that teams should no longer make a specific recommendation concerning leasing levels to the Secretary of the Interior. Instead, a set of alternatives would be developed by the team for secretarial decision. The governors of the western states, however, interpreted this Department change as an attempt to reduce their influence in federal coal-leasing decisions. In August nine western governors jointly wrote a letter to Interior Secretary James Watt complaining that "the effect of these changes is to once again centralize on the Potomac critical decisions affecting western states—decisions that should be made in the region."[11] Under continuing strong pressure from the western states, Secretary Watt in November 1982 agreed to reinstate the original policy that the regional coal team should make its own preferred leasing recommendation. He also agreed to provide a written statement of his reasons where he rejected the advice of a western governor concerning leasing in that governor's state.

The 1970s All Over Again?

By 1981 the federal coal program had achieved a degree of public acceptance that it had not known for many years. The western states were happy with the creation of the regional coal teams and the prospects for close federal-state coordination in planning federal coal leasing. Despite threats, environmental groups had not gone to court over the adequacy of the Interior Department EIS for the federal coal program or challenged the formal adoption of the program. The coal industry had complaints about some coal program details, but generally it looked forward to a resumption of federal coal leasing after many years of its absence.

Only two years later, however, public acceptance has faded rapidly. The complaints heard in the 1970s have reappeared with much of their old vigor. The Interior Department is said to be leasing too much coal, when there is no need to lease, given a large inventory of existing federal coal leases and insufficient demand to absorb this inventory. The Interior Department is said to be failing to achieve fair market value and to be giving away coal to private industry at bargain prices. Instead of the Interior Department, as in 1971, or the courts, as in 1977, Congress is now talking of imposing its own moratorium on future federal coal leasing. One cannot help but wonder whether the coal program history of the 1970s will be repeated once again.

Much the same groups are involved. The National Wildlife Federation has replaced the Natural Resources Defense Council as the most active environmental group opposing Interior Department leasing policies. As in the 1970s, environmentalists are supported by eastern and midwestern coal mining interests who also oppose western coal development. On April 26, 1983, Senator Alan Dixon of Illinois introduced a bill for a one-year moratorium on new federal coal leasing, bluntly commenting that "we have 5,600 unemployed coal miners in Illinois. . . . I cannot in good conscience . . . condone policies which do grievous harm to my state. I am sure that my colleagues from other coal-producing states share my concerns."[12]

During the 1970s, debates focused on the quantity of coal that needed to be leased. The price received was an important but less central issue. In the 1980s matters have been reversed. Achieving fair market value is now the focus of controversy and the quantity of coal leased an important but less central consideration. Whereas leasing moratoria of the 1970s were based primarily on an alleged failure to demonstrate the "need to lease," current proposed moratoria are based primarily on an alleged failure to demonstrate receipt of fair market value. Yet, for a demand or supply curve, one axis shows price and the other axis shows quantity; the price of new federal coal reserves implies a specific quantity of reserves leased, and vice versa. One fears that, possibly, the only coal-leasing change from the 1970s has been to switch the axis for debate.

Notes

Preface

1. For an account of the role of the office and of the work of a policy analyst, see Christopher K. Leman and Robert H. Nelson, "Ten Commandments for Policy Economists," *Journal of Policy Analysis and Management* 1 (Fall 1981): 97–117.

2. I served as the energy economist for the Office of Coal Leasing Planning and Coordination, and worked closely with the Department of Energy concerning computer modeling to calculate production goals for federal coal-leasing purposes. Other tasks included writing options papers and longer policy studies. I assisted in writing the "programmatic" environmental impact statement for the federal coal-management program that was released in April 1979.

3. Robert H. Nelson, *Zoning and Property Rights: An Analysis of the American System of Land Use Regulation* (Cambridge, Mass: MIT Press, 1977; paperback 1980).

4. Theodore J. Lowi, *The End of Liberalism: Ideology, Policy and the Crisis of Public Authority* (New York: W. W. Norton, 1969).

5. James Q. Wilson, "What Can be Done?" in *AEI Public Policy Papers* (Washington, D.C.: American Enterprise Institute, 1981), p. 19.

6. Robert H. Nelson, "Making Sense of the Sagebrush Rebellion" (paper prepared for presentation at the Third Annual Conference of the Association for Public Policy Analysis and Management, Washington, D.C., October 23–25, 1981); and Robert H. Nelson, "The Public Lands," in *Current Issues in Natural Resource Policy*, ed. Paul R. Portney (Washington, D.C.: Resources for the Future, distributed by Johns Hopkins University Press, 1982).

Introduction

1. For a classic description of a main part of the area, see Walter Prescott Webb, *The Great Plains* (Boston: Ginn, 1931).

2. For basic coal information, see The President's Commission on Coal, *Coal Data Book* (Washington, D.C.: Government Printing Office, 1980). See also Bureau of Land Management, U.S. Department of the Interior, *Final Environmental Statement, Federal Coal Management Program* (Washington, D.C.: Government Printing Office, April 1979), chap. 2.

3. For a classic economic study of coal use in its heyday as an energy source in England, see W. Stanley Jevons, *The Coal Question: An Inquiry Concerning the Progress of the Nation, and the Probable Exhaustion of Our Coal Mines* (New York: Augustus M. Kelley Reprints, 1965; repr. of 1906 ed.).

4. Leasing Policy Development Office, U.S. Department of Energy, *The 1980 Biennial Update of National and Regional Coal Production Goals for 1985, 1990 and 1995* (Washington, D.C.: DOE, Dec. 1980).

5. *Final Environmental Statement, Federal Coal Management Program*, p. 2–33.

6. For basic data on federal coal, see *Final Environmental Statement, Federal Coal Management Program*, chap. 2; also Congress of the United States, Office of Technology Assessment, *An Assessment of Development and Production Potential of Federal Coal Leases* (Washington, D.C.: Government Printing Office, Dec. 1981).

1. Conservationism and the Mineral Leasing Act of 1920

1. Samuel Trask Dana and Sally K. Fairfax, *Forest and Range Policy: Its Development in the United States* (New York: McGraw-Hill, 1980), p. 3.

2. See Robert W. Swenson, "Legal Aspects of Mineral Resources Exploitation," in *History of Public Land Law Development*, ed. Paul W. Gates, written for the Public Land Law Review Commission (Washington, D.C.: Government Printing Office, 1968), p. 724.

3. Samuel P. Hays, *Conservation and the Gospel of Efficiency: The Progressive Conservation Movement, 1890–1920* (Cambridge, Mass.: Harvard University Press, 1959), p. 82.

4. Bureau of Land Management, U.S. Department of the Interior, *Final Environmental Statement, Federal Coal Management Program* (Washington, D.C.: Government Printing Office, April 1979), pp. 2–10.

5. Richard Hofstadter, *The Age of Reform: From Bryan to F.D.R.* (New York: Vintage Books, 1955), pp. 10–11.

6. Woodrow Wilson, "The Study of Administration," *Political Science Quarterly* 2 (June 1887), repr. in *Ideas and Issues in Public Administration: A Book of Readings*, ed. Dwight Waldo (New York: McGraw-Hill, 1953), p. 71.

7. Dwight Waldo, *The Administrative State: A Study of the Political Theory of American Public Administration* (New York: Ronald Press, 1948), p. 19.

8. Ibid., pp. 67, 68, 80, 91.

9. Hays, *Conservation and the Gospel of Efficiency*, pp. 2–3, 265–66.

10. Ibid., p. 268.

11. Quoted in Ashley L. Schiff, *Fire and Water: Scientific Heresy in the Forest Service* (Cambridge, Mass.: Harvard University Press, 1962), p. 5.

12. Ibid., p. 165.

13. Gifford Pinchot, *Breaking New Ground* (New York: Harcourt, Brace, 1947), pp. 284, 260.

14. 51 *Cong. Rec.*, p. 14944 (1914), cited in Federal Trade Commission, *Report to the Federal Trade Commission on Federal Energy Land Policy: Efficiency, Revenue, and Competition* (Washington, D.C.: FTC, Oct. 1975, repr. by Senate Committee on Interior and Insular Affairs, 94th Cong., 2d Sess. 1976), p. 51.

15. 59th Cong., 2d Sess., *Sen. Doc. 310*, 1, cited in Hays, *Conservation and the Gospel of Efficiency*, p. 85.

16. 51 *Cong. Rec.*, p. 14944 (1914), cited in *Report to the Federal Trade Commission on Federal Energy Land Policy*, p. 52.

17. 58 *Cong. Rec.*, p. 4250 (1919), cited in *Report to the Federal Trade Commission on Federal Energy Land Policy*, p. 54.

18. 58 *Cong. Rec.*, p. 4253 (1919), cited in *Report to the Federal Trade Commission on Federal Energy Land Policy*, p. 55.

19. U. S. House of Representatives, Report No. 1138, 65 Cong., 3d Sess., p. 19 (1919), cited in *Report to the Federal Trade Commission on Federal Energy Land Policy*, pp. 57–58.

2. Congressional Intent and Opposite Results

1. The basic reference on the history of the public lands is Paul W. Gates, *History of Public Land Law Development*, written for the Public Land Law Review Commission (Washington, D.C.: Government Printing Office, 1968). See also Benjamin H. Hibbard, *A History of the Public Land Policies* (Madison: University of Wisconsin Press, 1965; original, 1924); Roy M. Robbins, *Our Landed Heritage: The Public Domain 1776–1970* (Lincoln: University of Nebraska Press, 1976; first ed., 1942); and E. Louise Peffer, *The Closing of the Public Domain: Disposal and Reservation Policies 1900–1950* (Stanford, Calif.: Stanford University Press, 1951).

2. Gates, *History of Public Land Law Development*, pp. 484, 486.

3. Report of the Public Lands Commission, 1880, 46th Cong., 2d Sess., *H. Ex. Doc.* No. 46, p. 9, quoted in Robbins, *Our Landed Heritage*, p. 290.

4. *The Western Range*, S. Doc., 74th Cong., 2d Sess. No. 199, 1936, p. 13. See Gates, *History of Public Land Law Development*, pp. 528–29.

5. Phillip O. Foss, *Politics and Grass: The Administration of Grazing on the Public Domain* (Seattle: University of Washington Press, 1960), p. 37.

6. Federal Trade Commission, *Report to the Federal Trade Commission on Federal Energy Land Policy: Efficiency, Revenue and Competition* (Washington, D.C.: FTC, repr. by Senate Committee on Interior and Insular Affairs, 94th Cong., 2d Sess. 1976), pp. 603–24. See also Council on Economic Priorities, *Leased and Lost: A Study of Public and Indian Coal Leasing in the West* (New York: CEP, 1974).

7. *Report to the Federal Trade Commission on Federal Energy Land Policy*, pp. 605, 609, 610, 615, 618, 622, 604, and 618.

8. Ibid., p. 608.

9. Gary Bennethum, "Holdings and Development of Federal Coal Leases" (Washington, D.C.: Bureau of Land Management, 1971), pp. 5, 8, and 11.

10. *North Central Power Study, Report of Phase I, Volume 1*, prepared under the direction of the Coordinating Committee, North Central Power Study (Oct. 1971), p. 5.

11. *Leased and Lost*, pp. 3, 12.

3. The Strong Opposition to Western Coal Development

1. John McPhee, *Encounters with the Archdruid* (New York: Farrar, Straus and Giroux, 1971), pp. 82–83.

2. Ibid., p. 83.

3. Hal Lindsey, with C. C. Carlson, *The Late Great Planet Earth* (Grand Rapids, Mich.: Zondervan Publishing House, 1970). Quote is from jacket of Nov. 1980 ed. (84th printing).

4. K. Ross Toole, *The Rape of the Great Plains: Northwest America, Cattle and Coal* (Boston: Little, Brown, 1976), p. 227.

5. Michael Parfit, *Last Stand at Rosebud Creek: Coal, Power, and People* (New York: E. P. Dutton, 1980), pp. 156–57.

6. Ibid., p. 165.

7. Quoted in Council on Economic Priorities, *Leased and Lost: A Study of Public and Indian Coal Leasing in the West* (New York: CEP, 1974), pp. 18–19.

8. Ibid., pp. 12, 7, 8, 13, 20, 18.

9. Roderick Nash, *Wilderness and the American Mind* (New Haven, Conn.: Yale University Press, 1973), chap. 10.

10. McPhee, *Encounters with the Archdruid*, pp. 164–65.

11. Katherine Fletcher, ed., *A Scientific and Policy Review of the Draft Environmental Impact Statement for the Proposed Federal Coal Leasing Program of the Bureau of Land Management, Department of the Interior*, (The Institute of Ecology, Environmental Impact Assessment Project, 1974).

12. Ibid., pp. 33, 92.

13. Ibid., pp. 49, 55–56, 66, 68–69.

14. Ibid., p. 69.

15. Ibid., pp. 24–25, 131.

16. Ibid., pp. 92, 27–28, 32.

17. Ibid., pp. 18, 15–16.

18. *Leased and Lost*, p. 20.

19. Ibid., pp. 3–4, 6, 9.

20. Ibid., pp. 6, 2, 29.

4. The Rejection of Central Planning and the Free Market

1. Marion Clawson, *The Bureau of Land Management* (New York: Praeger, 1971).

2. Bureau of Land Management, U.S. Department of the Interior, *Draft Environmental Impact Statement, Proposed Federal Coal Leasing Program, Volume 1* (Washington, D.C.: BLM, May 1974), pp. 1–2, 1–3.

3. Ibid., pp. 1–3, 1–4.

4. Memorandum from the Interior Assistant Secretary—Program Development and Budget (Roy Hughes) to the Undersecretary of the Interior (John Whitaker), "Coal Programmatic Statement," May 1, 1974, p. 1.

5. Bureau of Land Management, U.S. Department of the Interior, *Final Environmental Impact Statement, Proposed Federal Coal Leasing Program* (Washington, D.C.: Government Printing Office, Sept. 1975), pp. 9–25.

6. Ibid., pp. 9–69.

7. Contained in a famous letter by Secretary of Agriculture James Wilson but actually written by Pinchot; see Gifford Pinchot, *Breaking New Ground* (New York: Harcourt, Brace, 1947), p. 261.

8. Edward S. Mason, "Resources in the Past and for the Future," in *Resources for an Uncertain*

Future: Papers presented at a Forum marking the 25th anniversary of Resources for the Future, Oct. 13, 1977, Washington, D.C. (Baltimore: Johns Hopkins University Press for Resources for the Future, 1978), pp. 9–10.

9 . Marion Clawson, "The National Forests," *Science* 191 (Feb. 20, 1976): 766.

10. Marion Clawson, *The Economics of National Forest Management* (Washington, D.C.: Resources for the Future, June 1976), p. 84.

11. Anthony Downs, "Sustained Yield and American Social Goals" (paper presented at the Symposium, "The Economics of Sustained Yield Forestry," at the University of Washington, Seattle, Nov. 23, 1974). p. 25.

12. John V. Krutilla, "Adaptive Responses to Forces for Change" (paper presented at the Annual Meetings of the Society of American Foresters, Oct. 16, 1979, Boston, Mass.), p. 6.

13. "Public Land Coal Leasing Program," Attachment to Memorandum from Director, Bureau of Land Management (Curt Berklund) to Undersecretary of the Interior (John Whitaker), "Draft Coal Leasing Program Paper," Apr. 24, 1974, p. 6.

14. William Moffat, "(Draft) Policy Statement on Resumption of Federal Coal Leasing" (for proposed issuance by the Secretary of the Interior), Mar. 3, 1975, p. 13, Attachment to Memorandum from Deputy Undersecretary of the Interior (William Lyons) to Departmental Assistant Secretaries, "Coal Leasing," Mar. 26, 1975.

15. Memorandum from Director, Office of Minerals Policy Development (Darius Gaskins) to Assistant Secretary, Energy and Minerals (Jack Carlson), "Comments on Coal Leasing Federal Register Notice and Program Paper," May 14, 1974, p. 1.

16. Gunnar Myrdal, *Beyond the Welfare State: Economic Planning and Its International Implications* (New Haven, Conn.: Yale University Press, 1960), p. 19.

17. Alfred D. Chandler, Jr., *The Visible Hand: The Managerial Revolution in American Business* (Cambridge, Mass.: Harvard University Press, 1977).

18. See also Myrdal, *Beyond the Welfare State*, p. 19.

19. The classic statement is Harold Hotelling, "The Economics of Exhaustible Resources," *Journal of Political Economy*, 39 (Apr. 1931): 137–75. See also Robert M. Solow, "The Economics of Resources or the Resources of Economics," *American Economic Review* 64 (May 1974): 1–14.

20. Terry A. Ferrar, Barry L. Meyers, and George R. Neumann, *Legal and Economic Considerations in Federal Coal Leasing*, Report No. 15, Center for the Study of Environmental Policy, Pennsylvania State University, p. 35.

21. Memorandum from Assistant Secretary, Energy and Minerals (Jack Carlson), to Undersecretary of the Interior (John Whitaker), "Diligent Development Policy for Existing and New Coal Leases," Oct. 23, 1974, p. 1.

22. Council on Economic Priorities, *Leased and Lost: A Study of Public and Indian Coal Leasing in the West* (New York: CEP, 1974), p. 21.

23. U.S. Department of the Interior, "Option Paper, Diligence Requirements for Existing Coal Leases," undated, p. 3.

24. *Final Environmental Impact Statement, Proposed Federal Coal Leasing Program*, p. 1–9.

25. Letter from Mr. Ken Holum (General Manager, Western Fuels Association), to BLM Director Curt Berklund, Feb. 25, 1976, pp. 11–12.

26. Robert C. Anderson and Alison Silverstein, *Management of Federal Coal Properties in Areas of Fragmented Resource Ownership* (Washington, D.C.: Environmental Law Institute, Jan. 1980), pp. 2–28, 2–29.

27. Carl E. Bagge (President, National Coal Association), "Coal and the Public Lands and Why the Former Isn't Coming from the Latter When the Nation Needs It: A Case Study in National Masochism," (address delivered to the Rocky Mountain Energy-Minerals Conference, Billings, Mont., Oct. 15, 1975), p. 5.

28. Carlson memorandum on "Diligent Development Policy for Existing and New Coal Leases," p. 2.

5. A Planned Market Instead of a Free Market

1. Charles L. Schultze, *The Public Use of Private Interest* (Washington, D.C.: Brookings Institution, 1977), p. 2.

2. Ibid., pp. 5–6.

3. Lawrence R. Klein, *The Keynesian Revolution* (New York: Macmillan, 1961, 1st ed., 1947), pp. 153, 166.

4. See Allen V. Kneese and Charles L. Schultze, *Pollution, Prices and Public Policy* (Washington, D.C.: Brookings Institution, 1975) and Frederick R. Anderson et al., *Environmental Improvement Through Economic Incentives* (Baltimore: Johns Hopkins University Press for Resources for the Future, 1977).

5. For an excellent description of the current condition of the planned market and means of guiding it, see F. M. Scherer, *Industrial Market Structure and Economic Performance* (Chicago: Rand McNally, 1970).

6. Milton Friedman, *Capitalism and Freedom* (Chicago: University of Chicago Press, 1962).

7. Memorandum from Bill Moffat (Director of the Office of Policy Analysis) to Roy Hughes (Assistant Secretary for Program Development and Budget), "Do We Have a Coal Policy?" Jan. 24, 1975, p. 3.

8. Secretary of the Interior Thomas S. Kleppe, Interior Department press release, "New Federal Coal Leasing Policy to be Implemented Under Controlled Conditions," and accompanying "Statement on New Coal Leasing Policy," Jan. 26, 1976, p. 5.

9. Robert H. Nelson and Richard Winnor, "Options Paper on Diligent Development and Continuous Operation of Coal Leases" (Office of Policy Analysis, U.S. Department of the Interior, May 5, 1978), p. 7.

10. U.S. Department of the Interior, "Option Paper, Diligence Requirements for Existing Coal Leases" (undated), p. 6.

11. Ibid., p. 6

12. Letter from Marvin O. Young (Vice President, Peabody Coal Company) to the Director, Bureau of Land Management, Feb. 25, 1976, p. 3.

13. Bureau of Land Management, U.S. Department of the Interior, *Final Environmental Impact Statement, Proposed Federal Coal Leasing Program* (Washington, D.C.: Government Printing Office, Sept. 1975), p. I–13.

14. Secretary of the Interior Thomas Kleppe, Jan. 26, 1976 Press Release, p. 2.

15. Moffat Memorandum, "Do We Have a Coal Policy?" p. 3.

16. Ibid., p. 2.

17. William Moffat, "(Draft) Policy Statement on Resumption of Federal Coal Leasing" (for proposed issuance by the Secretary of the Interior), Mar. 3, 1975, p. 6, Attachment to Memorandum from Deputy Undersecretary (William Lyons) to Departmental Assistant Secretaries, "Coal Leasing," Mar. 26, 1975.

18. *Final Environmental Impact Statement, Proposed Federal Coal Leasing Program* (1975), p. I–13.

19. Moffatt memorandum, "Do We Have a Coal Policy?" p. 3.

20. William Moffat, "(Draft) Policy Statement on Resumption of Federal Coal Leasing," p. 5.

21. Federal Coal Management Program, U.S. Department of the Interior, *Final Report and Recommendations for the Secretary on Fair Market Value and Minimum Acceptable Bids for Federal Coal Leases* (Washington, D.C.: DOI, Dec. 1979), p. II–7.

22. See Donald J. Bieniewicz, *Improvements to the Federal Coal Leasing Program Linked to the Use of Intertract Bidding* (Washington, D.C.: Office of Policy Analysis, U.S. Department of the Interior, Apr. 1981); Donald J. Bieniewicz, *Federal Coal Leasing Targets and the Level of Leasing Issue—An Assessment (Part I)* (Washington, D.C.: Office of Policy Analysis, U.S. Department of the Interior, Oct. 1981); and Donald J. Bieniewicz, *Federal Coal Leasing Targets and the Level of Leasing Issue—An Assessment (Part II)* (Washington, D.C.: Office of Policy Analysis, U.S. Department of the Interior, Jan. 1982).

6. A New Economic Conservationism to Protect the Environment

1. Bureau of Land Management, U.S. Department of the Interior, *Final Environmental Impact Statement, Proposed Federal Coal Leasing Program* (Washington, D.C.: Government Printing Office, Sept. 1975), pp. 1–14.

2. See, for example, Robert Dorfman and Nancy S. Dorfman, eds., *Economics of the Environment: Selected Readings* (New York: W. W. Norton, 1972). For a good short exposition, see Larry E. Ruff, "The Economic Common Sense of Pollution," *The Public Interest* 19 (Spring 1970): 69–85, which is also included in the Dorfman collection.

3. A. C. Pigou, *The Economics of Welfare* (New York: Macmillan, 1932).

4. Ronald H. Coase, "The Problem of Social Cost," *Journal of Law and Economics* 3 (Oct. 1960): 1–44.

5. Memorandum from Assistant Secretary—Program Development and Budget (Roy Hughes) to the Undersecretary of the Interior, "Environmental Balancing in Mineral Leasing," Apr. 21, 1975, p. 4.

6. Office of Policy Analysis, U.S. Department of the Interior, "Decision Paper, The Definition and Application of the Terms 'Valuable Deposit' and 'Commercial Quantities'" (Washington, D.C., Dec. 1, 1975), p. 2.

7. Ibid., p. 8.

8. Congress of the United States, Office of Technology Assessment, *Management of Fuel and Non Fuel Minerals in Federal Lands: Current Status and Issues* (Washington, D.C.: OTA, Apr. 1979), pp. 226–27.

9. Ibid., p. 227.

10. Hughes memorandum on "Environmental Balancing in Mineral Leasing," p. 3.

11. Memorandum from Deputy Assistant Secretary—Energy and Minerals (Roland Reid) to the Secretary of the Interior, "Comments on the Terms 'Commercial Quantities' and 'Valuable Deposits,' as per Deputy Undersecretary's memorandum of June 20, 1975," June 27, 1975, p. 1.

12. Memorandum from the Solicitor (Kent Frizzell) to the Deputy Undersecretary (William Lyons), "Preference Right Leases—Coal, Sodium, Phosphate, Sulphur and Potassium," June 30, 1975, p. 3.

13. For another account of the commercial quantities decision, see Sally K. Fairfax and Barbara T. Andrews, "Debate Within and Debate Without: NEPA and the Redefinition of the 'Prudent Man' Rule," *Natural Resources Journal* 19 (July 1979): 505–35.

14. John V. Krutilla and Anthony C. Fisher, *The Economics of Natural Resource Environments: Studies in the Valuation of Commodity and Amenity Resources* (Baltimore: The Johns Hopkins University Press for Resources for the Future, 1975).

15. John V. Krutilla and John A. Haigh, "An Integrated Approach to National Forest Management," *Environmental Law* 8 (Winter 1978): 383.

16. Ibid., p. 399.

17. John V. Krutilla, "Adaptive Responses to Forces for Change" (paper presented at the Annual Meetings of the Society of American Foresters, Oct. 16, 1979, Boston, Mass.), p. 9.

18. Krutilla and Haigh, "An Integrated Approach to National Forest Management," p. 401.

19. Krutilla, "Adaptive Responses to Forces for Change," pp. 6, 11.

20. Darius W. Gaskins, Jr. (Acting Director, Office of Mineral Policy Development) and Richard A. Winnor (Policy Analyst, Office of Mineral Policy Development), "Policy Problems in Mineral Leasing" (unpublished paper, Washington, D.C., Oct. 1974). pp. 10–11.

21. Ibid., pp. 15–18.

22. Hughes memorandum on "Environmental Balancing in Mineral Leasing."

23. Ibid., pp. 1–2.

24. Office of Policy Analysis, "Decision Paper, The Definition and Application of the Terms 'Valuable Deposit' and 'Commercial Quantities,'" p. 5.

25. Ibid., p. 5.

26. Office of Mineral Policy Development, Department of the Interior, "Coal Leasing Goals" (Washington, D.C., 1974), p. 2.

7. The Judiciary Debates Its Proper Role in Federal Coal Planning

1. Charles M. Haar, "The Master Plan: An Impermanent Constitution," *Law and Contemporary Problems* 20 (Summer 1955): 353–418.

2. Allison Dunham, "Property, City Planning and Liberty," in *Law and Land, Anglo-American Planning Practice*, ed. Charles M. Haar (Cambridge, Mass.: Harvard University Press and MIT Press, 1964), p. 33.

3. Edward C. Banfield, "Ends and Means in Planning," *International Science Journal* 11 (1959): 364.

4. National Commission on Urban Problems, *Building the American City: Report of the National Commission on Urban Problems* (New York: Praeger, 1969), p. 223.

5. Herbert J. Gans, "The Need for Planners Trained in Policy Formulation," in *Urban Planning in Transition*, ed. Ernest Erber (New York: Grossman Publishers, 1970), pp. 243–44.

6. *Kleppe* v. *Sierra Club*, 427 U.S. at 421 (1976).

7. Ibid.

8. *Sierra Club* v. *Morton*, 514 F.2d at 866, 875 (1975).

9. Ibid., at 880, 883, 876–77.
10. Ibid., at 863.
11. Ibid., at 878, 874.
12. Ibid., at 871, 874.
13. Ibid., at 874.
14. Ibid., at 875.
15. Ibid., at 880.
16. William O. Douglas, *The Court Years: 1939–1945* (New York: Random House, 1980), p. 8.
17. *Sierra Club* v. *Morton*, at 884.
18. *Kleppe* v. *Sierra Club*, 427 U.S. 390 (1976). (The name of the case was changed from *Sierra* v. *Morton* due to the appointment of Thomas Kleppe as Secretary of the Interior in place of former Secretary Rogers C. B. Morton.)
19. Ibid., at 411, 412.
20. Ibid., at 409, 410, 414.
21. Ibid., at 406.
22. For a good discussion of recent trends in the judicial role, see Richard B. Stewart, "The Reformation of American Administrative Law," *Harvard Law Review* 88 (June 1975): 1667–1813; and, in response, Abram Chayes, "The Role of the Judge in Public Law Litigation," *Harvard Law Review* 89 (May 1976): 1281–1316.

8. The Interior Department Tries to Resume Leasing

1. Bureau of Land Management, U.S. Department of the Interior, *Final Environmental Impact Statement, Proposed Federal Coal Leasing Program* (Washington, D.C.: Government Printing Office, Sept. 1975). p. I–9.
2. Letter from NRDC staff members John D. Leshy and Terry R. Lash to Interior Secretary Rogers C. B. Morton, Feb. 3, 1975.
3. Memorandum from John Whitaker (Undersecretary of the Interior) to Frank Zarb, "ERC Policy Review, Federal Coal Leasing," Dec. 20, 1974, p. 4.
4. Ibid., p. 5.
5. General Accounting Office, *Role of Federal Coal Resources in Meeting National Energy Goals Needs to Be Determined and the Leasing Process Improved* (Washington, D.C.: GAO, 1976), pp. 26–27.
6. Statement of Katherine Fletcher on Federal Coal Leasing and Mining, before the House Subcommittee on Mines and Mining, House Interior Committee, representing the Environmental Defense Fund, Apr. 5, 1976, pp. 3, 5.
7. Ibid., p. 5.
8. Memorandum from Bill Moffat (Director, Office of Policy Analysis) to Roy Hughes (Assistant Secretary—Program Development and Budget), "Do We Have a Coal Policy?" Jan. 24, 1975, p. 4.
9. Memorandum from Assistant Secretary—Land and Water Resources (Jack Horton) to the Secretary of the Interior, "Coal Programmatic Environmental Impact Statement," June 30, 1975, pp. 1–3.
10. Lester G. Chase, *A Tabulation of City Planning Commissions in the United States* (Washington, D.C.: Bureau of Standards, Department of Commerce, 1931), p. 4. Available at Avery Library, Columbia University.

9. The Judiciary Mandates Central Planning

1. "The Environment: The President's Message to the Congress," May 23, 1977, repr. in Council on Environmental Quality, *Environmental Quality: The Eighth Annual Report of the Council on Environmental Quality* (Washington, D.C.: Government Printing Office, 1977), pp. 346, 352.
2. Office of the Assistant Secretary for Land and Water Resources, U.S. Department of the Interior, *Federal Coal Management Review* (Washington, D.C., July 21, 1977), p. 1.
3. Remarks of Assistant Secretary for Land and Water Resources Guy Martin, reported in Jerry Brown, "Coal Leasing: Back to the Drawing Board," *The Energy Daily* (Washington, D.C., June 21, 1977), p. 3.
4. Brown, "Coal Leasing: Back to the Drawing Board," pp. 2–3, 1.

5. For the full story of this policy-making episode, see Bruce A. Ackerman and William T. Hassler, *Clean Coal/Dirty Air: How the Clean Air Act Became a Multibillion-Dollar Bail-Out to High-Sulfur Coal Producers and What Should Be Done About It* (New Haven, Conn.: Yale University Press, 1981).

6. *Federal Coal Management Review*, p. 5.

7. Ibid., p. 9.

8. "Federal Defendant's Memorandum in Response to the Court's July 5, 1977 Request," Interior Department Submission to the United States District Court, District of Columbia, in connection with *NRDC* v. *Hughes*, July 25, 1977, p. 3.

9. Brown, "Coal Leasing: Back to the Drawing Board," p. 1.

10. *NRDC* v. *Hughes*, 437 F. Suppl. at 994 (1977).

11. Ibid., at 993.

12. Ibid., at 988.

13. Ibid., at 992.

14. Ibid., at 989.

15. Ibid., at 989, 991.

16. William Moffat, "(Draft) Policy Statement on Resumption of Federal Coal Leasing" (for proposed issuance by the Secretary of the Interior), Mar. 3, 1975, p. 6, Attachment to Memorandum from Deputy Undersecretary (William Lyons) to Departmental Assistant Secretaries, "Coal Leasing," Mar. 26, 1975.

17. Letter from Ed Herschler, Governor of Wyoming, to Interior Secretary Cecil Andrus, Jan. 27, 1978, p. 2.

18. Barry Commoner, *The Closing Circle: Nature, Man and Technology* (New York: Bantam Books, 1972), pp. 285–86.

19. *NRDC* v. *Hughes*, at 992.

10. A Lesser Role for the Coal Industry

1. Memorandum from Robert Nelson (Office of Assistant Secretary, Policy, Budget and Administration) to Deputy Assistant Secretary, Energy and Minerals (Charles Eddy), "Issues for Coal Policy Review," Aug. 22, 1977, p. 1.

2. U.S. Department of the Interior, "Options Paper: Departmental Approach for the Long-Term Coal Leasing Program" (Washington, D.C.: Department of the Interior, Oct. 7, 1977). p. 1.

3. Ibid., p. 4.

4. Ibid., pp. 6, 8, 9.

5. U.S. Department of the Interior, *Coal Data Task Force: Final Report* (Washington, D.C.: DOI, May 7, 1979), p. 23.

6. Ibid., p. 24.

7. Decision of Interior Secretary Cecil D. Andrus, "Long-Term Coal Leasing Options Decision Sheet," Oct. 22, 1977.

8. Memorandum from Joe Browder (staff assistant to Assistant Secretary, Land and Water Resources) to Chuck Eddy (Deputy Assistant Secretary, Energy and Minerals), "Options Paper on Leasing Alternatives," Sept. 22, 1977, pp. 1–2.

9. Ibid., p. 2.

10. Samuel P. Hays, *Conservation and the Gospel of Efficiency: The Progressive Conservation Movement, 1890–1920* (Cambridge, Mass.: Harvard University Press, 1959), p. 266.

11. Ibid., pp. 268–69.

11. Central Planning in Operation

1. General Accounting Office, *Role of Federal Coal Resources in Meeting National Energy Goals Needs to Be Determined and the Leasing Process Improved* (Washington, D.C.: GAO, 1976). p. 27.

2. Letter from Bruce J. Terris (NRDC Counsel) to George Turcott (Acting Director, BLM) containing "Comments of the Natural Resources Defense Council Concerning the Final Environmental Impact Statement of the Department of the Interior on the Proposed Federal Coal Leasing Program," Feb. 2, 1978, pp. 5–6.

3. "Memorandum of Understanding Between the Department of the Interior and the Department

of Energy Concerning the Establishment and Use of Production Goals for Energy Resources on Federal Lands," 1978, p. 1, repr. in Bureau of Land Management, U.S. Department of the Interior, *Final Environmental Statement, Federal Coal Management Program* (Washington, D.C.: Government Printing Office, Apr. 1979), p. B-1.

4. Remarks of Assistant Secretary of the Interior for Land and Water Resources, Guy Martin, before the Coal Week Symposium, Washington, D.C., Sept. 13, 1977, pp. 4–5.

5. Leasing Policy Development Office, U.S. Department of Energy, *Federal Coal Leasing and 1985 and 1990 Regional Coal Production Forecasts* (Washington, D.C.: DOE, June 1978); Leasing Policy Development Office, U.S. Department of Energy, *Working Paper: Interim Updates to 1985 and 1990 Regional Forecasts* (Washington, D.C.: DOE, Apr. 1979); and Leasing Policy Development Office, U.S. Department of Energy, *The 1980 Biennial Update of National and Regional Coal Production Goals for 1985, 1990 and 1995* (Washington, D.C.: DOE, Dec. 1980).

6. ICF Inc., *Final Report: The Demand for Western Coal and Its Sensitivity to Key Uncertainties*, prepared for the Departments of Interior and Energy (Washington, D.C.: Sept. 1978), p. 11.

7. For a general examination of forecasting problems, see William Ascher, *Forecasting: An Appraisal for Policy-Makers and Planners* (Baltimore: Johns Hopkins University Press, 1978), and J. Scott Armstrong, "The Seer-Sucker Theory: The Value of Experts in Forecasting," *Technology Review* 83 (June/July 1980): 18–24. Armstrong offers similar conclusions concerning expert capabilities.

12. Setting Coal-Leasing Targets

1. Bureau of Land Management, U.S. Department of the Interior, *Final Environmental Statement, Federal Coal Management Program* (Washington, D.C.: Government Printing Office, April 1979), p. 2–56.

2. Ibid., pp. 2–50, 2–56, 2–57.

3. Leasing Policy Development Office, U.S. Department of Energy, *Federal Coal Leasing and 1985 and 1990 Regional Coal Production Forecasts* (Washington, D.C.: DOE, June 1978), and Federal Energy Administration, "Summer Quarter Western Coal Development Monitoring System" (Washington, D.C.: FEA, Aug. 1, 1977); see also Office of Coal Leasing Planning and Coordination, U.S. Department of the Interior, *Background Paper—Levels of Leasing* (Washington, D.C.: DOI, June 26, 1978), p. 16.

4. Sherry H. Olson, *The Depletion Myth: A History of Railroad Use of Timber* (Cambridge, Mass: Harvard University Press, 1971), p. 179.

5. *Final Environmental Statement, Federal Coal Management Program*, p. 2–58.

6. Ibid., pp. 2–57, 2–58.

7. Ibid., p. 2–64.

8. Antitrust Division, U.S. Department of Justice, *Competition in the Coal Industry* (Washington, D.C.: DOJ, 1978), quoted in *Final Environmental Statement, Federal Coal Management Program*, p. 2–61.

9. National Coal Association, *Detailed Comments of the National Coal Association Upon Draft Environmental Statement, Federal Coal Management Program and Statement of Policy, "Coordination of Federal Land Review"* (Washington, D.C.: NCA, Feb. 1979). pp. 33–34, repr. in *Final Environmental Statement, Federal Coal Management Program*, p. K–134.

10. Letter from Charles W. Margolf to BLM Office of Coal Management, Feb. 7, 1979, pp. 3–4, repr. in *Final Environmental Statement, Federal Coal Management Program*, p. K–6.

11. *Comments of the Natural Resources Defense Council, Inc., on the Draft Environmental Statement, for the Proposed Federal Coal Management Program of the Department of the Interior*, prepared by Johanna Wald and Jonathan Lash, pp. 21–22, 28, repr. in *Final Environmental Statement, Federal Coal Management Program*, pp. K–75, K–77.

12. Ibid., p. K–78.

13. Ibid., p. K–78.

14. *Comments on Draft Environmental Statement, Federal Coal Management Program, by Friends of the Earth*, prepared by David C. Masselli, Kevin L. Markey, and John Weiner, Feb. 13, 1979, pp. 9–10, repr. in *Final Environmental Statement, Federal Coal Management Program*, p. K–105.

15. Ibid., p. K–105.

16. Ibid., p. K–104.

17. Ibid., p. K–104.

18. U.S. Department of the Interior, *Secretarial Issue Document, Federal Coal Management*

Program, Volume II–Need for Leasing Program Implementation (Washington, D.C.: DOI, May 1979), Executive Summary, p. 5.

19. "The Future of Federal Coal" (speech by Steven P. Quarles, Deputy Undersecretary, Department of the Interior, at the American Mining Congress 1979 Mining Convention, Los Angeles, Calif., Sept. 25, 1979), p. 10.

20. Ibid., pp. 10–11.

21. Ibid., p. 10.

22. Testimony of Donald L. Flexner, Deputy Assistant Attorney General, Antitrust Division, before the Subcommittee on Mines and Mining of the House Committee on Interior and Insular Affairs, Concerning Federal Coal Leasing, June 25, 1979, pp. 5–6.

23. Letter from Donald L. Flexner, Deputy Assistant Attorney General, Antitrust Division, to Frank Gregg, Director, Bureau of Land Management, pp. 10–11.

24. Attachment 1 to letter from Ruth M. Davis (Assistant Secretary, Resource Applications, Department of Energy) to Guy Martin (Assistant Secretary, Land and Water Resources, Department of the Interior), Oct. 7, 1980, p. 4 of Attachment; and General Accounting Office, *A Shortfall in Leasing Coal from Federal Lands: What Effect on National Energy Goals* (Washington, D.C.: GAO, Aug. 22, 1980), pp. 27, 21.

13. The Advance and Retreat of Land-Use Planning

1. For discussions of land-use planning on the public lands, see *One Third of the Nation's Land: A Report to the President and to the Congress by the Public Land Law Review Commission* (Washington, D.C.: Government Printing Office, 1970); Glen O. Robinson, *The Forest Service: A Study in Public Land Management* (Baltimore: The Johns Hopkins University Press for Resources for the Future, 1975); Benjamin W. Hahn, J. Douglas Post, and Charles B. White, *National Forest Management: A Handbook for Public Input and Review* (Stanford, Calif.: Stanford Environmental Law Society, Sept. 1978); and Christopher K. Leman, *Resource Assessment and Program Development: An Evaluation of Forest Service Experience Under the Resources Planning Act, with Lessons for other Natural Resources Agencies* (Washington, D.C.: Office of Policy Analysis, Department of the Interior, Aug. 1980).

2. See, for example, John Delafons, *Land-Use Controls in the United States* (Cambridge, Mass.: Joint Center for Urban Studies of the Massachusetts Institute of Technology and Harvard University, 1962); Richard Babcock, *The Zoning Game, Municipal Practices and Policies* (Madison: University of Wisconsin Press, 1966); and Robert H. Nelson, *Zoning and Property Rights: An Analysis of the American System of Land Use Regulation* (Cambridge, Mass.: The MIT Press, 1977; paperback, 1980).

3. For examinations of the broad discretion in the term "multiple use," see John A. Zivnuska, "The Multiple Problems of Multiple Use," *Journal of Forestry* (Aug. 1961); George R. Hall, "The Myth and Reality of Multiple Use Forestry," *Natural Resources Journal* 3 (Oct. 1963): 276–90; Charles A. Reich, "A Precious Resource," *The Center Magazine* (Jan./Feb. 1975), repr. of 1962 original; R. W. Behan, "The Succotash Syndrome, or Multiple Use: A Heartfelt Approach to Forest Land Management," *Natural Resources Journal* 7 (Oct. 1967): 473–84; E. M. Sterling, "The Myth of Multiple Use," *American Forests* 72 (June 1970): 24–27; and R. W. Behan, "Forestry and the End of Innocence," *American Forests* 81 (May 1975): 16–19, 38–49.

4. Sterling, "The Myth of Multiple Use," quoted in Richard M. Alston, *Forest Goals and Decision Making in the Forest Service* (Ogden, Utah: Intermountain Forest and Range Experiment Station, U.S. Forest Service, Sept. 1972), p. 2.

5. Bureau of Land Management, U.S. Department of the Interior, *Final Environmental Impact Statement, Proposed Federal Coal Leasing Program* (Washington, D.C.: Government Printing Office, Sept. 1975). p. I–14.

6. Statement of Katherine Fletcher on Federal Coal Leasing and Mining, before the House Subcommittee on Mines and Mining, House Interior Committee (for the Environmental Defense Fund), Apr. 5, 1976, pp. 2–3.

7. Office of the Assistant Secretary, Land and Water Resources, U.S. Department of the Interior, *Final Report, Coal Task Force 2, Land Unsuitability Criteria* (Washington, D.C.: DOI, Sept. 11, 1978), p. 12.

8. Ibid., p. 12.

9. Ibid., p. ii.

10. Ibid., pp. 8–9, 11.
11. Ibid., p. viii.
12. Ibid., p. 12.
13. *Detailed Comments of the National Coal Association Upon Draft Environmental Statement, Federal Coal Management Program, and Statement of Policy, "Coordination of Federal Land Review,"* repr. in Bureau of Land Management, U.S. Department of the Interior, *Final Environmental Statement, Federal Coal Management Program* (Washington, D.C.: Government Printing Office, Apr. 1979), p. K–128.
14. Ibid., p. K–129.
15. *Final Report, Coal Task Force 2, Land Unsuitability Criteria*, p. 9.
16. Ibid., pp. 9–10.
17. Ibid., pp. 10–11.
18. Ibid., p. viii.
19. Ibid., p. x.
20. U.S. Department of the Interior, *Coal Data Task Force: Final Report*, (Washington, D.C.: DOI, May 7, 1979), pp. 2–3.
21. Ibid., p. 8.
22. Ibid., pp. 22–23.
23. John Leonard Watson, "The Federal Coal Follies–A New Program Ends (Begins) a Decade of Anxiety??" *Denver Law Journal*, 58 (1980): 132. The quote within the quote is from *National Coal Policy Project, Mining Task Force–Coal Leasing Group, Land Use Planning and Market Forces in Federal Coal Management, Fourth Discussion Draft* (Feb. 12, 1980)
24. See, for example, Ernest Erber, ed., *Urban Planning in Transition* (New York: Grossman Publishers, 1970), and John Friedmann and Barclay Hudson, "Knowledge and Action: A Guide to Planning Theory," *AIP Journal* 40 (Jan. 1974): 2–16. For broader criticisms of formal planning, see the writings of Charles E. Lindblom, most recently, Charles E. Lindblom and David K. Cohen, *Usable Knowledge: Social Science and Social Problem Solving* (New Haven, Conn.: Yale University Press, 1979). For a useful business critique of planning, see George A. Steiner, *Strategic Planning: What Every Manager Must Know* (New York: Free Press, 1979).
25. For a good discussion of such issues, see Peter G. W. Keen and Michael S. Scott Morton, *Decision Support Systems: An Organizational Perspective* (Reading, Mass.: Addison-Wesley, 1978).

14. Interest-Group Liberalism in Practice

1. Herbert Simon, "The Proverbs of Administration," *Public Administration Review* (Winter 1946): 53–67.
2. Dwight Waldo, *The Administrative State: A Study of the Political Theory of American Public Administration* (New York: Ronald Press, 1948), pp. 177, 128.
3. John Kenneth Galbraith, *American Capitalism: The Concept of Countervailing Power* (Boston: Houghton Mifflin, 1956), and John Kenneth Galbraith, *The New Industrial State* (Boston: Houghton Mifflin, 1967).
4. Gunnar Myrdal, *Beyond the Welfare State: Economic Planning and Its International Implications* (New Haven, Conn.: Yale University Press, 1960), p. 96.
5. David B. Truman, *The Governmental Process: Political Interests and Public Opinion* (New York: Alfred A. Knopf, 1951), pp. 50–51.
6. Ibid., p. 443.
7. Pendleton Herring, "Research on Government, Politics, and Administration," in *Research for Public Policy*, Brookings Dedication Lectures (Washington, D.C.: Brookings Institution, 1961), p. 13.
8. Theodore J. Lowi, *The End of Liberalism: Ideology, Policy and the Crisis of Public Authority* (New York: W. W. Norton, 1969), pp. 47, 46.
9. Myrdal, *Beyond the Welfare State*, pp. 46–47.
10. Ibid., p. 116.
11. "The Environment: The President's Message to the Congress," May 23, 1977, repr. in Council on Environmental Quality, *Environmental Quality: The Eighth Annual Report of the Council on Environmental Quality* (Washington, D.C.: Government Printing Office, 1977). p. 352.
12. Remarks of Assistant Secretary for Land and Water Resources Guy Martin, reported in Jerry Brown, "Coal Leasing: Back to the Drawing Board," *The Energy Daily* (Washington, D.C., June 21, 1977), p. 2.

13. Remarks of Assistant Secretary of the Interior Guy Martin, before the Coal Week Symposium, Washington, D.C., Sept. 13, 1977, p. 3.

14. Ibid., pp. 3, 2.

15. Department of the Interior News Release, "Andrus Announces New Federal Coal Management Program," June 4, 1979, p. 1.

16. Department of the Interior News Release, "Andrus Announces First Sale Under Interior's New Coal Leasing Program," Oct. 20, 1980, p. 1.

17. Letter to the editor from John D. Leshy (former Associate Solicitor of the Interior Department for Energy and Resources) to *The Arizona Republic*, Phoenix, Ariz., Jan. 10, 1981.

18. Remarks of Steven P. Quarles (Deputy Undersecretary of the Interior), reported in Christopher Madison, "Leasing Federal Lands for Coal–When Is Enough Really Enough?" *National Journal* (Washington, D.C., Nov. 15, 1980), p. 1942.

19. John Leonard Watson, "The Federal Coal Follies–A New Program Ends (Begins) a Decade of Anxiety??" *Denver Law Journal* 58 (1980): 133.

20. Madison, "Leasing Federal Lands for Coal–When Is Enough Really Enough?" p. 1942.

21. Remarks of the Honorable Cecil D. Andrus, Secretary of the Interior, to the Fourth Annual Coal Conference and Exposition, Louisville, Ky., Oct. 18, 1977, p. 14.

22. Option paper on "Need for Leasing/Leasing Systems Choice," pp. 10, 12. Contained in Office of Coal Leasing Planning and Coordination, U.S. Department of the Interior, *Secretarial Issue Paper: Formulation of Proposal for Coal Programmatic Environmental Impact Statement* (Washington, D.C.: DOI, June 23, 1978; available in Interior Department library).

23. *Federal Coal Management Report: Fiscal Year 1979*, Annual Report of the Secretary of the Interior (Washington, D.C.: Government Printing Office, Mar. 1980), p. 16.

24. 43 CFR 3420.3, published in *The Federal Register* 44 (July 19, 1979): 42619.

25. Ibid., pp. 42619, 42620.

26. Watson, "The Federal Coal Follies," p. 134.

27. John V. Krutilla and Anthony C. Fisher, with Richard E. Rice, *Economic and Fiscal Impacts of Coal Development: Northern Great Plains* (Baltimore: The Johns Hopkins University Press for Resources for the Future, 1978), pp. 185, 177, 180.

28. Memorandum from the Secretary of the Interior to the President, "Sixty Day Coal Report," June 4, 1979, p. 2.

29. Friedrich A. Hayek, *The Road to Serfdom* (Chicago: University of Chicago Press, 1944), pp. 66–67.

30. Lowi, *The End of Liberalism*, p. 101.

15. The Rediscovery of the Planned Market

1. Remarks of Jack Campbell (Council on Wage and Price Stability), reported in Christopher Madison, "Leasing Federal Lands for Coal–When Is Enough Really Enough?" *National Journal* (Nov. 15, 1980), p. 1938.

2. Warren Johnson, *Muddling Toward Frugality* (San Francisco: Sierra Club Books, 1978), p. 130.

3. Friedrich A. Hayek, *The Constitution of Liberty* (Chicago: Henry Regnery Company, 1960), p. 400.

4. Johnson, *Muddling Toward Frugality*, pp. 130–31.

5. P.L. 94–377, 90 Stat. 1083 (1976), Sec. 3 (3) (C).

6. "Floor Statements on S. 391 by Senator Metcalf," repr. in *Federal Coal Leasing Policies and Regulations*, report of the Senate Committee on Energy and Natural Resources, No. 95–77, 95th Cong., 2d Sess., 1978, pp. 116–17.

7. Attachment in letter from Bruce J. Terris (NRDC Counsel) to George L. Turcott (Acting Director, BLM), "Comments of the Natural Resources Defense Council Concerning the Final Environmental Impact Statement of the Department of the Interior on the Proposed Federal Coal Leasing Program," Feb. 2, 1978, p. 13.

8. William Watson and Richard Bernknopf, *Economic Analysis of Maximum Economic Recovery of Federal Coal* (Washington, D.C.: Program Analysis Office, U.S. Geological Survey, Department of the Interior, May 29, 1979), p. 1.

9. "Option Paper on Maximum Economic Recovery," in Office of Coal Leasing Planning and Coordination, U.S. Department of the Interior, *Secretarial Issue Paper: Formulation of Proposal for Coal Programmatic Environmental Impact Statement* (Washington, D.C.: DOI, June 23, 1978).

10. Watson and Bernknopf, *Economic Analysis of Maximum Economic Recovery of Federal Coal*, pp. 2–3.

11. Memorandum from the Secretary of the Interior to the President, "Sixty Day Coal Report," June 4, 1979, p. 5.

12. Council on Economic Priorities, *Leased and Lost: A Study of Public and Indian Coal Leasing in the West* (New York: CEP, 1974), p. 22.

13. Remarks of Robert Uram (Associate Solicitor, U.S. Department of the Interior), in U.S. Department of the Interior, *Proceedings of the Town Meeting on Fair Market Value of Federal Coal Leases*, held in Denver, Col., Nov. 1, 1979 (Washington, D.C.: DOI, 1980), pp. 137–38.

14. Interior Department News Release on "New Federal Coal Leasing Policy to Be Implemented Under Controlled Conditions," Jan. 26, 1976, p. 1.

15. P.L. 94–377, 90 Stat. 1083 (1976), Sec. 2(1).

16. Federal Coal Management Program, U.S. Department of the Interior, *Final Report and Recommendations for the Secretary on Fair Market Value and Minimum Acceptable Bids for Federal Coal Leases* (Washington, D.C.: DOI, Dec. 1979), pp. II–6, II–7.

17. Ibid., pp. II–8, II–7.

18. Ibid., p. II–4.

19. Ibid., p. II–4.

20. Ibid., p. I–5.

21. Ibid., p. I–10.

22. Richard L. Gordon, "Problems of Fair Market Value Estimation," presented to the Fair Market Value Task Force, Department of the Interior, p. 5; included in ICF, *(Draft) Observations on Fair Market Value* (Washington, D.C.: ICF, Nov. 1979).

23. Richard L. Gordon, "Fair Market Value Estimation in the Context of Coal Leasing Policies," presented to the Fair Market Value Task Force, Department of the Interior, p. 3; included in ICF, *(Draft) Observations on Fair Market Value* (Washington, D.C.: ICF, Nov. 1979).

24. Remarks of Professor James B. Ramsey (Chairman, Department of Economics, New York University), in *Proceedings of the Town Meeting on Fair Market Value of Federal Coal Leases*, pp. 59, 109–10, 59–60.

25. Remarks of Laurence I. Moss, in *Proceedings of the Town Meeting on Fair Market Value of Federal Coal Leases*, pp. 198, 85.

26. Richard A. Clark, Robert C. Lind, and Robert Smiley, *Enhancing Competition for Federal Coal Leases*, prepared for the Office of Policy Analysis, U.S. Department of the Interior (McLean, Va.: Science Application, Inc., Jan. 1976), p. 57.

27. Robert C. Anderson and Alison Silverstein, *Management of Federal Coal Properties in Areas of Fragmented Resource Ownership* (Washington, D.C.: Environmental Law Institute, Jan. 1980), pp. 1, 2 of Recommendations.

28. Office of Policy Analysis, U.S. Department of the Interior, "Bidding Systems for Coal Leasing" (Washington, D.C.: DOI, Feb. 2, 1976); Thomas Teisberg and Robert H. Nelson, "Coal Tract Selection and Bidding System Option Paper" (Washington, D.C.: Office of Policy Analysis, Department of the Interior, May 5, 1978); and C. B. McGuire, "Intertract Competition and the Design of Lease Sales for Western Coal Lands" (paper presented to the 53rd Annual Western Economic Association Conference, Hawaii, June 25, 1978).

29. Federal Trade Commission, *Report to the Federal Trade Commission on Federal Energy Land Policy: Effficiency, Revenue and Competition* (Washington, D.C.: FTC, Oct. 1975), esp. chaps. 3, 5, and 12.

30. One of the few available studies was Wallace E. Tyner and Robert J. Kalter, *Western Coal: Problems or Promise* (Lexington, Mass.: Lexington Books, 1978). Also, Stephen L. McDonald, *The Leasing of Federal Lands for Fossil Fuels* (Baltimore: Johns Hopkins University Press for Resources for the Future, 1979); and, most recently, Richard L. Gordon, *Federal Coal Leasing Policy: Competition in the Energy Industries* (Washington, D.C.: American Enterprise Institute for Public Policy Research, 1981).

16. Lessons in Political Economy

1. Alfred D. Chandler, Jr., *The Visible Hand: The Managerial Revolution in American Business* (Cambridge, Mass.: Harvard University Press, 1977).

2. Peter F. Drucker, *The Practice of Management* (New York: Harper & Row, 1959), pp. 41–42. See also Peter F. Drucker, *Concept of the Corporation* (Boston: Beacon Press, 1960), original ed., 1946.

3. Samuel P. Hays, *Conservation and the Gospel of Efficiency: The Progressive Conservation Movement, 1890–1920* (Cambridge, Mass.: Harvard University Press, 1959).

4. Dwight Waldo, *The Administrative State: A Study of the Political Theory of American Public Administration* (New York: Ronald Press, 1948), p. 66.

5. Wayne A. Leeman, ed., *Capitalism, Market Socialism and Central Planning: Readings in Comparative Economic Systems* (Boston: Houghton Mifflin, 1963); and Oskar Lange and Fred M. Taylor, *On the Economic Theory of Socialism* (Minneapolis: University of Minnesota Press, 1938).

6. C. A. R. Crosland, *The Future of Socialism* (New York: Schocken Books, 1963), original ed., 1956; Michael Harrington, *Socialism* (New York: Bantam Books, 1973); and Charles E. Lindblom, *Politics and Markets: The World's Political-Economic Systems* (New York: Basic Books, 1977).

7. Michael Ellman, *Socialist Planning* (Cambridge: Cambridge University Press, 1979). An earlier leading work is Abram Bergson, *The Economics of Soviet Planning* (New Haven, Conn.: Yale University Press, 1964).

8. Ellman, *Socialist Planning*, pp. 17–18.

9. Ibid., pp. 65–66.

10. Ibid., pp. 66–67, 72–73.

11. Ibid., p. 71.

12. James Q. Wilson, *Political Organizations* (New York: Basic Books, 1973).

13. Guy Paul Land, "Judicial Process and the Decline of Twentieth-Century American Liberalism," *Harvard Journal of Legislation* 16 (Spring 1979): 286.

14. "A Way Out of the Wilderness," editorial commentary in *Marysville* (Calif.) *Appeal-Democrat*, Mar. 11, 1982.

15. Terry L. Anderson and Peter J. Hill, *The Birth of the Transfer Society* (Stanford, Calif.: Hoover Institution Press, 1980).

16. Murray N. Rothbard, *For a New Liberty: The Libertarian Manifesto* (New York: Collier Books, 1973).

17. John Baden and Richard L. Stroup, eds., *Bureaucracy vs. Environment: The Environmental Costs of Bureaucratic Governance* (Ann Arbor: University of Michigan Press, 1981).

17. The Future of Federal Coal

1. Charles E. Lindblom, "The Science of 'Muddling Through,'" *Public Administration Review* 19 (Spring 1959); David Braybrooke and Charles E. Lindblom, *A Strategy of Decision: Policy Evaluation as a Social Process* (New York: Free Press, 1963); Charles E. Lindblom, *The Intelligence of a Democracy: Decision Making Through Mutual Adjustment* (New York: Free Press, 1965); Charles E. Lindblom and David K. Cohen, *Usable Knowledge: Social Science and Social Science Problem Solving* (New Haven, Conn.: Yale University Press, 1979); Aaron B. Wildavsky, *The Politics of the Budgetary Process* (Boston: Little, Brown, 1964); and Aaron B. Wildavsky, *Speaking Truth to Power: The Art and Craft of Policy Analysis* (Boston: Little, Brown, 1979).

2. Robert H. Nelson, "An Analysis of 1978 Revenues and Costs of Public Land Management by the Interior Department in 13 Western States" (Office of Policy Analysis, U.S. Department of the Interior, Dec. 1979).

3. Richard Nehring, Benjamin Zycher, with Joseph Wharton, *Coal Development and Government Regulation in the Northern Great Plains: A Preliminary Report*, Rand Corporation Report R-1981-NSF/RC (Santa Monica, Calif.: Rand Corporation, Aug. 1976), p. 148.

4. Arthur M. Okun, *Equality and Efficiency: The Big Tradeoff* (Washington, D.C.: Brookings Institution, 1975), p. 51. See also Charles L. Schultze, *The Public Use of Private Interest* (Washington, D.C.: Brookings Institution, 1977).

5. See Robert H. Nelson, "The Public Lands," in *Current Issues in Natural Resource Policy*, Paul R. Portnoy, ed. (Washington, D.C.: Resources for the Future, distributed by Johns Hopkins University Press, 1982).

Epilogue

1. See U.S. Department of the Interior, *Federal Coal Management Report, Fiscal Year 1981*, annual report of the Secretary of the Interior under Section 8 of the Federal Coal Leasing Amendments Act of 1976 (Washington, D.C.: Government Printing Office, 1982).

2. *A Report to the Committee on Appropriations, U.S. House of Representatives, on the Coal Leasing Program of the U.S. Department of the Interior*, prepared by the Surveys and Investigations Staff, House Appropriations Committee (Washington, D.C., April 1983).

3. United States General Accounting Office, *Analysis of the Powder River Basin Federal Coal Lease Sale: Economic Valuation Improvements and Legislative Changes Needed*, a report to the Congress by the Comptroller General (Washington, D.C.: GAO, May 11, 1983).

4. Letter from Garrey E. Carruthers (Interior Assistant Secretary for Land and Water Resources) to Representative Sidney R. Yates (Chairman, House Subcommittee on Interior and Related Agency Appropriations), and attachments, May 17, 1983.

5. *Analysis of the Powder River Basin Federal Coal Lease Sale*, p. iii.

6. "Memorandum to All Members of the House Appropriations Committee," from James Watt, Secretary of the Interior, May 17, 1983.

7. Revised coal management program regulations (43 CFR 3400) were published in July 1982. See 47 F.R. 33114, July 30, 1982.

8. Letter from Donald Paul Hodel (Under Secretary of the Interior) to Senator Henry M. Jackson, and attached questions and answers, July 19, 1982. See Jackson question 19 and Interior response (p. 15 of attachment).

9. Ibid., Jackson question 11 and Interior response (pp. 8-9 of attachment).

10. Ibid., Interior response to Jackson question 9 (p. 8 of attachment).

11. Letter from Governor Ed Herschler of Wyoming and eight other western governors to James Watt, Secretary of the Interior, Aug. 30, 1982, p. 1.

12. *Congressional Record*—Senate, p. S5213, April 26, 1983.

Index

Robert H. Nelson is a member of the Economics Staff, Office of Policy Analysis, U.S. Department of the Interior. He is author of *Zoning and Property Rights*.